ASSET

DATA

INTEGRITY

IS SERIOUS
BUSINESS

Robert S. DiStefano

Stephen J. Thomas

Industrial Press Inc.

New York

Library of Congress Cataloging-in-Publication Data

DiStefano, Robert S.
  Asset data integrity is serious business / Robert S. DiStefano, Stephen J.
Thomas.
     p. cm.
  Includes bibliographical references and index.
  ISBN 978-0-8311-3422-8 (hbk.)
  1. Production management. 2. Plant performance--Statistics--Evaluation.
3. Information technology--Management. 4. Quality control--Management. 5.
Industrial management. I. Thomas, Stephen J. II. Title.

  TS155.D576 2010
  658.2'7028557--dc22

                              2010035592

    Industrial Press, Inc.
    989 Avenue of the Americas
    New York, NY 10018

Sponsoring Editor: John Carleo
Developmental Editor: Robert Weinstein
Interior Text and Cover Design: Janet Romano

                  10   9   8   7   6   5   4   3   2   1

Dedication

To all of my colleagues at MRG
                    -Bob DiStefano

To Jonah Aden Thomas, my grandson
                    -Steve Thomas

Table of Contents

Acknowledgements

I would like to thank all of my colleagues at MRG who made it possible for me to dedicate the time necessary to write my chapters for this book. I'd especially like to acknowledge Bruce Hawkins, Jason Price, Brian Maier, Steve Cyr, Chris Jensen, Ken Bass, Tom Daley, Dennis Belanger and Wes Graf for their contributions of valuable content, in some cases drafts of entire chapters. Without their invaluable help, I never would have been able to complete my part of this project. Lastly, I would like to thank my dear friend and co-author Steve Thomas, who asked me to assist him in creating this work and who made it a unique and rewarding experience.

-Bob DiStefano

For many years I had the privilege of leading a data integrity team. These individuals, functional experts in the various areas of maintenance and reliability, worked full time cleaning up the existing data and putting work processes in place to sustain their effort. The work they did, and the problems they overcame, convinced me beyond a shadow of doubt that there was a need for a text on this often overlooked subject. They deserve recognition, both collectively and individually, for their behind-the-scenes, often unrecognized efforts on this very important topic. The team consisted of Janet Bailey, Erin Darrow, Paul Gatti, Phil Giannini, Amy Grove, Tony LaRocca, Mike Russo, Wayne Smith, and Clyde Tinklepaugh.

-Steve Thomas

Together, we would like to acknowledge John Carleo and Patrick Hansard of Industrial Press Inc. They are more than just our contacts with the publishing company. They have been great advisors and supporters for this very important work.

We would also like to thank Janet Romano of Industrial Press for her creativity in developing a truly great cover. She also took the Word documents and artwork and assembled them into a very nicely formatted book. With multiple authors, this was no easy task.

Our thanks are also extended to Robert Weinstein, our manuscript editor. Robert took our writing and fixed the problems so that you, the reader, can enjoy and benefit from this book. He has a wonderful knack for fixing things but never altering the concepts or ideas in our original text.

About the Authors

Robert S. DiStefano is the Chairman and CEO of Management Resources Group, Inc., a professional services firm specializing in reliability engineering and asset data integrity. With more than 30 years of professional engineering, maintenance, reliability, management, and consulting experience, Bob specializes in advising and coaching executives in leveraging reliability engineering and master data management principles into tangible business benefits at the enterprise level. He is an active member of the enterprise asset management expert community and has held board positions in SMRP, including Director of the Body of Knowledge. Bob has authored numerous articles and is often a featured speaker at industry forums. He holds degrees in engineering and management and is a Certified Maintenance and Reliability Professional.

Stephen J. Thomas has spent 40 years working in the petrochemical industry. Through personal involvement at all levels of the maintenance and reliability work process, he has gained extraordinarily broad experience in all phases of strategic and tactical development and implementation of organizational change. Steve has a B.S. in Electrical Engineering from Drexel University and M.S. degrees in Systems Engineering and Organizational Dynamics from the University of Pennsylvania. He has written five books focusing on change management, including *Improving Maintenance and Reliability Through Cultural Change*, all available from Industrial Press.

Introduction

The integrity of our plant asset data, one of the most overlooked items on the business landscape, has the potential to deliver a significant contribution to your company's bottom line. If your asset data is not reliable, you need to convince the organization of the enormous potential that is locked away. To accomplish this, you need to understand the breadth of the problem and the value of solving it. A viable business case for action is needed—so let's get started!

Chapter 1

The Business Case for Data Integrity

In God we trust
all others, bring data!

Unknown

1.1　Introduction to the Business Case

A vast and growing amount of data is being accumulated in businesses today. Yet many people in business intuitively know there is not a corresponding improvement in reliable information on which to base good decisions. In fact, in many cases just the opposite is happening—despite more and more data, finding information that can be trusted is increasingly difficult.

Let's be honest. To many people, no business subject is more dull than the subject of "data." Nevertheless, the subject of data integrity is written about in business journals more often than many other seemingly more interesting topics. Furthermore, many surveys reveal a growing concern among business executives related to the ability to take advantage of the reams of data that are being collecting.

If reading about data makes your eyes glaze over, we wrote this book for you. We've attempted to put the seemingly dull subject of data integrity in an interesting context and tried to make it come alive.

1

Although we were excited about the subject of data before (call us unusual!), we are more excited now as a result of having written this text. We hope that reading this text will help you become excited as well. Equally important, we hope you will find yourself in a better position to be able to do something about data integrity in your company.

Your intuition may tell you that there are large benefits associated with bringing integrity to your business data. We must admit, however, that intuition is not enough to garner the proper level of senior management support and resources to improve the data. You need a convincing business case whose development can prove to be very challenging for several reasons. First and foremost, the business case for Data Integrity is so vast, so far-reaching, so all-encompassing, and so pervasive in every aspect of business that knowing where to start, and how much of the story to tell, is a daunting proposition. We think the best approach is to frame the case in broad terms, citing specific facts and some quantitative examples that support the intuition that the business case for Data Integrity is huge. Armed with this information, you should then be able to personalize the case for data integrity in your firm or plant.

1.2   Information Overload

Consider this: the average installed data storage capacity at Fortune 1000 corporations has grown from 198 terabytes to 680 terabytes in less than two years. This is growth of more than 340% and capacity continues to double every ten months! That statistic puts into objective terms what we all instinctively know about our data—we have huge quantities of it and we are accumulating more and more every day.

1.3 **Searching for Data**

But what else do we know about our data? In an article in *Information Week* (January 2007), writer Marianne Kolbasuck McGee reported the results of a study conducted by Accenture which surveyed 1,009 managers from U.S.- and U.K.-based companies with annual revenue of more than $500 million. The study found that average middle managers spend about two hours a day looking for data they need. The study does not comment on how often the search ends successfully. But we can assume that at least some of that time is wasted. Why? Several reasons.

First, the volume of data is too large and most of it is not needed. To arrive at the needed data, one has to cull through reams of irrelevant or unnecessary data. As the CEO of Business Objects, Bernard Lieutaud has said, "There is too much data and it is duplicated hundreds of times. The mistake companies make is that they start from the data they have. They need to ask what data their users need and what questions are the users asking."

Second, the quality of the data—or data integrity—is generally poor. Much of the data is inaccurate, out of date, inconsistent, incomplete, poorly formatted, or subject to interpretation. As Andy Bitterer, an analyst with Gartner, has said, "There's not one company that doesn't have a data quality problem. Most companies have about 200 data sources and much of it is poor quality and inconsistent." Therefore, even when you do arrive at the needed data, can you trust it? If you have to hesitate to answer that question, you're undoubtedly spending some time deciding whether the data you have finally found (assuming you actually found it) is trustworthy and whether you can rely on it to accomplish your task at hand.

There are other reasons, but these two alone are compelling. Let's try to quantify these phenomena. The U.S. Department of Labor's Bureau of Labor Statistics indicates that in May 2006 approximately 142 million workers were in the U.S. workforce. Assume conservatively that only 10% of those workers are middle managers, as defined in the Accenture study mentioned above. Also assume conservatively that only 25% of the two hours per day spent searching for data is wasted (many studies indicate the actual percentage is higher). The amount is 1,633,000,000 hours (that's right … billion!) that are wasted annually in the United States alone—equivalent to about 785,000 man-years annually!

To put these figures in financial terms, suppose $40 per hour is the average loaded cost rate for middle managers. Then $65,320,000,000 is wasted every year—that's $65.32 billion annually, just in the United States! (When we tried to do the math on an accountant's desktop calculator, the final result was an error. We had to open an Excel spreadsheet to perform the calculations!!) Imagine what this number is when calculated worldwide!

Can we put those 1,633,200,000 freed-up hours per year (or 785,000 workers per year) to good use? Most certainly!

1.4   Retiring Baby Boomers

Assuming we can fix the data integrity problem nationwide and free up these hours, some of the retiring workers won't have to be replaced. Thus, the cost structure of the company will go down. According to the U.S. Department of Labor's Bureau of Labor Statistics, approximately 22.8 million people aged 55 and older are in the U.S. workforce today—approximately 16% of the entire workforce. (By the way, the number of workers in this category is growing four times faster than the workforce as a whole!)

Assume conservatively that the number of workers in this category does *not* increase. Further assume that the 22.8 million of them will retire evenly over the next ten years. In short, approximately 2.3 million workers will retire each year in the United States (the actual estimated number is higher). The freed-up hours related to data integrity could account for about a third of that. Thus one-third of those retired workers would not have to be replaced—assuming we solve the data integrity problem.

Again, our assumptions in this example are conservative. It is entirely possible that simply fixing the data integrity problems could go a long way toward solving the aging / retiring workforce debacle in the United States and elsewhere.

This analysis deals strictly with an efficiency gain. We have not yet talked about the effectiveness of our efforts or, to put it another way, the impact of the "Brain Drain" on the knowledge residing inside the corporation.

1.5   The Brain Drain

More than 80% of U.S. manufacturers face a shortage of qualified craft workers. This shortage is because of the retiring workforce phenomenon, and the fact that fewer new workers are entering the skilled trades, or even technical degree programs. As a result, we don't have a feedstock of replacement workers ample enough—or skilled enough—to replace the retirees.

This challenge should put the onus on management of our industrial companies to figure out not only how to leverage a potentially smaller workforce through eliminating wasted activities. It should also challenge them, perhaps more importantly, to institutionalize and memorialize the knowledge currently in the heads of these workers in the company's systems and data sources. Wouldn't meeting this challenge head-on facilitate and accelerate the accumulation of skills and knowledge on the part of new less-skilled workers?

In addition, the institutionalization of knowledge and information could facilitate the same work being satisfactorily accomplished by less-skilled workers. In other words, it is possible that we won't have to completely replace the retiring workers in-kind. The combination of better systems, automation, information, procedures, guidelines, training media, etc. (all of which have "data" at their heart)—with less-skilled workers—could represent a game-changing step change in how we go about doing the work in our manufacturing and industrial companies! That step change could permanently and favorably impact the cost of doing business.

1.6   A Business Case Example

Let's take the data just presented and provide an example closer to home for the average industrial plant worker.

Many studies in the maintenance and reliability (or physical asset management) field, including several conducted by Management Resources Group, Inc., have pointed consistently to an estimate that between 30 and 45 minutes per day per maintenance worker is wasted searching for spare parts because of poor catalog data integrity. Spare parts represent just one narrow area of the many aspects of physical asset management, but they provide a helpful example. (Incidentally, according to research presented in *Maintenance Planning and Scheduling Handbook* by Richard (Doc) Palmer, the total amount of unproductive time on the part of an industrial maintenance worker is, on average, 5 hours and 45 minutes per day! That means that productive time, on average, is only 28%! Not all of that unproductive time is related to data integrity, but some of it certainly is.)

If you are not familiar with this aspect of physical asset management, be aware that inventory catalog descriptions are generally not formatted consistently or in a way that facilitates rapid searching and finding of the needed spare part. Searchers often become frustrated because they cannot easily find the part in question. Sometimes the time spent searching doesn't even result in a successful find, let alone a rapid one. Typical problems include: the system indicates that the needed part is in stock, but when visiting the bin there are actually none in stock; the indicated bin location is wrong; the searcher spelled a search word differently than the myriad of ways it exists in the catalog material master records (e.g., Bearing, BRG, Brg).

Referring to the U.S. Department of Labor's Bureau of Labor Statistics May 2006 Occupational Employment and Wage Estimates, it is estimated that the United States has approximately 5.45 million industrial maintenance workers today. The same data indicates that the mean hourly wage rate for these workers is approximately $20. A loaded cost including fringe benefits would be approximately $26 per hour (assuming a 30% adder for fringe benefits).

If each of these workers is wasting conservatively 30 minutes per day searching for spare parts, then we are wasting 626,750,000 hours per year in the United States. That's over 300,000 workers per year, or 5% of the industrial maintenance workforce. At the mean loaded cost per hour that equates to $16,295,200,000 annually—$16.295 billion!

Are we suggesting that the primary manifestation of these potential gains is a reduction in head count? Not necessarily, although the natural attrition generated by the Baby Boomers' retirements will present opportunities to reduce head count without having to lay off any workers.

In addition, you gain the real opportunity to redeploy the freed-up resources to more value-added activities that will drive higher equipment reliability and lower maintenance costs. The consensus of the expert community in asset management is that most industrial plants rely too heavily on time-based preventive maintenance (PM) procedures as a primary maintenance strategy. Based on the results of thousands of PM Optimization initiatives, approximately 60% of existing preventive maintenance activities in existence are inappropriate strategies for the assets in question. Thus, a very large portion of the maintenance workforce is engaged in low-value or zero-value work. Expert

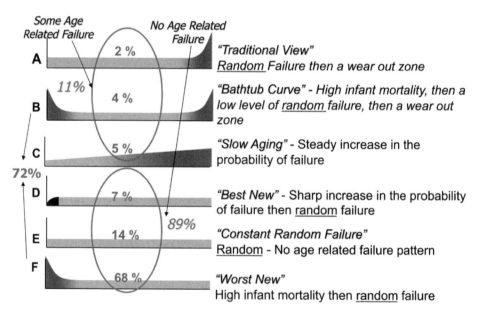

Figure 1-1 Asset Failure Behavior Patterns

(printed with permission of Industrial Press, Inc.)

analysis of equipment failure behavior, using proven tools like Reliability Centered Maintenance (RCM) and Failure Modes and Effects Analysis (FMEA), dictates that the vast majority of assets in a typical industrial complex—about 89%—do **not** observe a predictable time-based failure pattern. Only about 11% of assets do so, as Figure 1-1 clearly shows.

The failure curves depicted in Figure 1-1 are accepted and proven knowledge dating back to studies that began in the 1960s. Keep in mind that the curves in this figure show the probability of failure on the basis of time (the x-axis is time in these curves). What these curves tell us is that it is impossible to predict failures of 89% of the assets in a plant *on the basis of time*. That does not mean we cannot predict failure for these classes of assets—it simply means that we cannot do so on the basis of time.

If the failure behavior of a specific class of asset shows that the asset fails randomly on the basis of time, how can we accurately define an interval for preventive, or time-based, maintenance? We can't! Yet that is exactly what we have tried to do for the past fifty years. Typically we have guessed what the correct and safe time interval should be for preventive maintenance based on the actual historical failure behavior of that asset.

Consider an asset that over a five-year period ran for 1 year before its first failure, then after repair it ran for 6 months before its next failure, then 3 months, then 18 months, then 5 months, then 16 months. What time interval would we set to do preventive maintenance on this asset if we wanted to prevent failure? If the asset is critical to operations, we'd have to take a risk-conservative approach and say that we should do something to this asset every 3 months. (Remember, two of the runs during the five-year period were less than 6 months).

But based on the actual five-year failure history of this asset, that would mean that we would have done preventive maintenance much too often. Remember the machine usually ran without trouble for longer than 3 months during this five-year period. If we were doing PM on that asset quarterly, we would have done PM work unnecessarily 14 times during that five-year period!

Not only did the machine not need a PM during many of those runs, but as we can see from Figure 1-1, we may have introduced defects that actually induced failures that otherwise would not have occurred. This phenomenon is referred to in the reliability profession as Infant Mortality. Many people have probably heard the phrase "if it ain't broke, don't fix it." Well, this adage has more merit than you would think.

As you can see in Figure 1-2, there is significant basis and proof, dating back to the 1960s, to support elimination of many existing PMs. This elimination would free up significant manpower, potentially to be used to hedge against the loss of knowledge with retiring Baby Boomers, or it can be used to redeploy some of the freed-up resources to perform other more value-added tasks that would be required to enhance asset performance.

Most equipment does not observe a time-based failure pattern, Therefore, should we do no maintenance at all on the 89% of assets and simply wait for them to fail? Absolutely not. In fact, while we cannot predict failure for these assets on the basis of *time*, we most certainly *can* predict failure of these assets on the basis of *condition*— using a variety of sensitive technologies and tools designed to detect early warnings of impending failures. These sensitive technologies and tools are commonly referred to as *Predictive Maintenance and Condition Monitoring*. Examples of such tools include vibration analy-

Other Studies of Failure Patterns

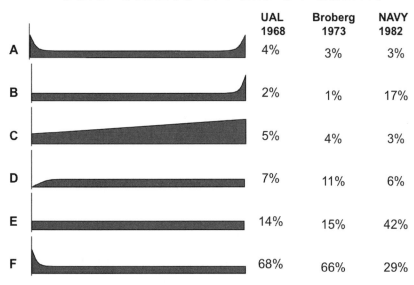

	UAL 1968	Broberg 1973	NAVY 1982
A	4%	3%	3%
B	2%	1%	17%
C	5%	4%	3%
D	7%	11%	6%
E	14%	15%	42%
F	68%	66%	29%

Percentage of items exhibiting failure rate pattern

Figure 1-2 Studies Showing Asset Failure Behaviors

sis (used on rotating assets), infrared thermography (used on electrical distribution equipment and other asset classes), oil analysis (used on lubricated equipment), and ultrasonic inspection (used on stationary equipment, piping and valves). There are others that we don't need to go into here.

The trick to PM optimization (reduction) and proper deployment of predictive maintenance tools is first to know how to categorize the assets, using analysis methods designed to understand likely and costly failure modes. Then with that knowledge, review the existing preventive maintenance procedures. Eliminate those that are either not addressing failure modes or are applied to asset types that don't observe any time-based pattern. Once these steps are undertaken, the appropriate Predictive Maintenance strategies must be deployed. The result of this optimization of the maintenance program invariably is a significant reduction in work, with the attendant reduction in labor and spare parts usage. In turn, these results drive significant cost savings and enhanced asset performance.

You may be asking yourself at this stage, "What does this all have to do with data integrity?" Well, how can you possibly accomplish this optimization if your foundational data sources lack integrity and quality—i.e., are incomplete, inconsistent, and inaccurate? If you don't have an accurate and complete equipment list, for example, you lack a fundamental prerequisite to unlocking these technical benefits. The answer is that, without asset data integrity, you cannot accomplish the optimization described here, particularly if you want to do so both efficiently and effectively. Let's explore this a bit more.

1.7 Consistency or Lack Thereof

Most corporations have allowed different industrial plants in the company's asset fleet significant autonomy in choice and use of systems, formatting of master foundational data in those systems, maintenance strategies, etc. It is typical today that multiple plants in one corporation have similar, if not identical, assets. Yet these assets are described differently from plant to plant. The maintenance strategies that are deployed for these assets vary dramatically from plant to plant. Wide variation in maintenance strategy across a fleet of like assets results in a corresponding variation in the operating performance of these similar assets. Some assets operate more reliably whereas other assets of similar or identical class operate unreliably.

Based on our knowledge of best practices, why would we allow this in any company? Wouldn't we want to use sound analytical methods to classify our assets, analyze their failure modes, and apply somewhat consistent maintenance strategies across the enterprise (taking into consideration some differences are warranted given operating context, etc.)? It seems logical and makes common sense to want to do so. But how can we undertake these steps efficiently if our assets are not described with a consistent taxonomy across the enterprise? Once again, we can't.

For those who may not be familiar with the term "taxonomy," it refers to the system of classification that guides the consistent formatting and nomenclature assignment used to describe whatever is being classified. A consistent taxonomy allows you to identify the like assets across the fleet and then measure and solve for the variation. An inconsistent taxonomy seriously impairs your ability to optimize your asset maintenance strategies and achieve consistent, reliable operation of your assets across the fleet. At the most basic level, this is a data integrity issue that must be solved in order to tap into the potential cost savings and improved asset performance that are waiting to be unlocked. Without data integrity, a significant entitlement of business benefits is locked away and unattainable.

1.8   The Data Integrity Corporate Entitlement

You might ask how big is this entitlement or opportunity in cost savings and improved asset performance benefits. Is it worth investing in addressing the data integrity issues? We will try to quantify those benefits in general, but objective terms next.

Benchmark statistics from various industries reliably measure the range of maintenance spending in various industries. Some such benchmarking systems include the Solomon Associates' benchmarking in the petroleum industry, the Townsend benchmarks in the polymers industry, the cross-industry benchmarks produced by the Society for Maintenance and Reliability Professionals, and the cross-industry benchmark system of Management Resources Group, Inc. These various sources all indicate that the average industrial plant annually spends between 5% and 10% of the Replacement Asset Value (RAV) on maintenance. (These expenses include maintenance labor, spare parts costs, contracted maintenance-related services, and condition monitoring expenses.) Top quartile plants (best-in-class) spend less than 3%, and top decile plants spend as little as 1.5% on maintenance annually. Interestingly, fourth quartile performers (the poorest performers) spend as much as 15% of RAV annually! This spread is significant.

Another interesting fact is found as these benchmarks are applied to multiple plants within a single corporation. That application usually reveals a fairly wide variation of performance—some plants are good, and some have room for improvement. This fact is rooted primarily in the autonomy traditionally granted individual plants. Varying maintenance strategies on similar equipment contribute more than any other factor to varying performance levels of that similar equipment when comparing plant to plant. The ultimate goal of any industrial corporation should be to unlock the entire entitlement across the enterprise. They should solve for these variations and, to whatever reasonable extent possible, standardize the maintenance practices across the fleet—taking care to ensure that the chosen strategies are in fact the ones that will drive the best performance.

Using the maintenance cost as a percentage of the Replacement Asset Value metric, we can estimate the excess maintenance spending in the United States as follows:

According to statistics published in 2007 by the U.S. Department of Commerce (namely the Current-Cost Net Stock of Private Fixed Assets), the estimated Replacement Asset Value of industrial plants throughout the fifty United States is well over $5 trillion.

We applied the average benchmarks of maintenance spending cited earlier. We assumed every plant in the United States could reach top quartile performance from where they currently stand (which would mean they would all spend annually 3% or less of Replacement Asset Value on maintenance—and have high equipment reliability). We then calculated potential annual savings of $233 billion, based strictly on reduction of maintenance expenses. It is extremely important to caution here that simply cutting maintenance budgets does not result in top-quartile performance. In fact, that mistake has been made during the past ten years or so. Because maintenance practices were not changed, equipment condition degraded. Today, many plants are in a reactionary and, therefore, expensive maintenance mode as a result of those mistakes. This is supported by numerous surveys including some done by the Aberdeen Group (Boston, MA). As discussed earlier, the cost of maintenance will go down as a lagging result of changing the maintenance practices—e.g., analysis of the equipment failure behavior, categorization of the assets, optimization/reduction of the time-based Preventive Maintenance activities, and introduction of condition-based Predictive Maintenance-driven activities.

The expert community in asset management uses a reliable and widely-adopted metric to estimate potential operational benefits of improved maintenance practices. This metric was initially established by Rohm and Haas during its successful work improving operational excellence across its chemical plant fleet in the 1980s. This metric holds that the operational benefits of best maintenance practices will typically equate to between three and seven times the maintenance cost reduction. Thus, for every dollar of reduced maintenance spending, an additional three to seven dollars accrue as operational benefits.

Operational benefits include less unscheduled downtime, higher first-run product quality, better delivery punctuality, lower energy consumption, and better regulatory compliance. If we use the conservative three times multiplier, then another $699 billion of benefits are available—conservatively estimated. This amount makes the total potential business case for best asset management practices in the United States $932 billion annually! Estimates of the world-wide opportunity range from $2 to $4 trillion annually.

Once again, this significant benefit cannot be unlocked without a solid foundation of accurate, complete, and consistently-formatted technical data (data integrity).

1.9   Impact on Shareholder Value

This discussion of a corporate entitlement would not be complete without attempting to put that entitlement into terms that are meaningful to senior executives in an industrial company, as if dollars and cents are not senior executive terms. What is the most important language spoken at that level? If you said "money" you are partially right, because financial benefits are always extremely important. But what do the large potential benefits mentioned above really mean to your CEO? Shareholder or stakeholder value is usually the most important attribute motivating the senior executives—and that is not purely financial.

Shareholder or stakeholder value is enhanced as a result of many outcomes of better data integrity, including better regulatory compliance, more sustainable (fuel efficient and environmentally friendly) operations, lower unit cost of goods, better delivery punctuality, higher product quality, lower working capital needs, and lower inventories. All of these attributes will enhance the value of an enterprise. The full entitlement of all of these attributes, and others, cannot be fully achieved without data integrity.

So how do these significant financial benefits impact shareholder value? The U.S. Department of Commerce's Bureau of Economic Analysis tracks Industrial Production on an annual basis. Industrial Production is the sum of all of the sales booked by enterprises in all industries in the United States. If a breakfast cereal company has annual sales of $100 billion, but they purchase $25 billion of wheat from an agricultural company supplier, the value of the $25 billion sales of the agricultural company's wheat sold to the cereal company is added to Industrial Production. (As an aside, Gross Domestic Product, or GDP, would not double count the sales of the wheat. It would only consider the value added for all of this activity. Therefore, GDP is almost always considerably less than Industrial Production). Economics lesson aside, U.S. Industrial Production in 2007 was $25.8 trillion. That is the sum of all of the revenue booked by all of the industrial companies in the United States.

Let's pretend that one single public company generated all of the $25.8 trillion in revenue. Let's further assume that this fictitious company has a net profit margin of 10% (or $2.58 trillion in profit), a Price to Earnings ratio of 6:1, and 25 billion shares of stock outstanding. Earlier in this chapter we conservatively identified almost $820 billion in potential benefits associated with three areas of data integrity: time wasted by middle managers searching through data, time wasted by hourly industrial maintenance workers searching for spare parts, and the cost savings and operational benefits associated with proper deployment of standardized best asset management practices. What happens to shareholder value of this fictitious company if that $820 billion in benefits is actually realized? Refer to Figure 1-3.

Income Statement	Before	After	Notes
Sales (millions)	$25,809,000.	$25,809,000.	
Total Cost of Doing Business (millions)	$23,228,100.	$22,408,199.	$820 Billion (Savings)
Net Profit (millions)	$2,580,900.	$3,400,900.	
Profit Margin (% of Sales)	10%	13.2%	31.8% Increase in Profit
Price to Earnings Ratio	6:1	6:1	
Market Capitalization (Value) of Company (millions)	$15,485,400	$20,405,400	31.8% Increase in Market Capital
Shares Outstanding	25,000,000,000	25,000,000,000	
Earnings per Share (EPS)	$103.24	$136.04	31.8% Increase in EPS

Figure 1-3 Shareholder Value Impact of Data Integrity Benefits

Notice in Figure 1-3 that the profit margin of this company goes from 10% to 13.2%. That change is a nearly 32% increase in profit! Notice also that the market capitalization, or value of this company, goes from about $15.5 trillion to about $20.4 trillion—again, a nearly 32% increase. Earnings per share likewise increases almost 32%—an increase of $32.80 in earnings per share, with the stock share price going from $108 to $136.

Any CEO would consider this impact on shareholder value to be strategic and game-changing, by any measure.

What is the potential impact in your company?

PART ONE

Understanding the Importance
of
Asset Data Integrity

Chapter 1 makes a powerful statement about the potential value of a properly-managed data integrity effort. To be able to address the opportunity, you need a deep understanding of everything that is involved in bringing integrity to your asset data and supporting it over time. Part One will provide that understanding and start you on a journey that will deliver your untapped business value.

Chapter 2

Plant Asset Information — A Keystone for Success

> For the want of shoe, a horse was lost
> For the want of a horse, a rider was lost
> For the want of a rider, a battle was lost
> For the want of a battle, a kingdom was lost
>
> *Unknown*

2.1   Overview

It is an irrefutable fact that without properly-managed, asset-related data, your business could be in trouble. This data applied at the enterprise level and every other level in the organizational hierarchy is critically important to your success and ultimately your very survival!

We all recognize that there is a vast amount of data available that relates to our plant assets. When properly combined, these data elements yield asset-related information, which is a vitally important part of every business. When properly managed and applied in support of business decisions, the potential exists to drastically improve your return on investment—and, of course, the converse is equally true.

Consider what your largest investment is in any plant. The answer is obviously your fixed assets, which in most cases are worth very large amounts of money. When these assets are performing at an optimum level, the potential clearly exists for you to be able to maximize your return on your assets, often called ROA.

Let's examine what prevents us from maximizing this return. Begin by excluding those impacts over which we have limited control: external supply and demand. This limits our boundary for the analysis to the physical plant and those assets that take the raw material and convert it to finished or semi-finished goods. Analysis as to why we can't optimize our return on assets falls into four categories: 1) people and work process-related issues, 2) machinery and its lack of reliability, 3) methods associated with production and maintenance of the equipment, and 4) materials management. Each of these areas all depend on one thing which, if properly managed, will set the stage for optimum performance and maximized ROA—and that is asset-related data.

We live in what has been called the "Information Age." Our businesses generate so much data that, as you read in chapter one, storage requirements are doubling every nine months. However, data at this granular level of basic information is useless if not properly managed and employed to create maximum return.

How is it possible that something so small and often overlooked— the data about plant assets—has the potential for such a major, long-term impact on the business at the enterprise level? Consider the following scenarios:

A company decided to overhaul an existing processing line within their plant. An engineering firm was hired to design and oversee the construction. At the end of the project, a large amount of asset-related data was delivered to the owner both electronically and in several massive information manuals. Of course, the owner didn't have the time to manage the data provided, so the electronic media were stored in the company's computer system; meanwhile the data books found a home on a shelf in Engineering in case anyone needed the information.

Why does this matter? This "data dump" is often the very manner in which asset information is delivered at the end of a project. What possible impact could this have on the business? Actually, it doesn't matter, not until those at the site need the asset information to run the process, perform maintenance on the equipment, or make any decisions related to optimizing production—an everyday occurrence in industry. In the condition that the data was delivered and in the manner in which the company stored it, retrieval by the site personnel for their use will be extremely difficult and time consuming. The effect of this at the plant level can be devastating. Wrong decisions can easily be made, opportunities lost, and possibly even safety or environmental protocols compromised. For lack of accurate data, the business can indeed suffer at the plant level.

Now let's look at a scenario that involves the highest level of the organizational hierarchy: the CEO. Accurate, reliable, and easily accessible data about plant assets create valuable information needed by the middle management level tasked with operating the plant. This information also provides the basic needs of the senior levels to conduct rigorous analysis and make proper decisions which ultimately will directly contribute to return on the assets employed in the manufacturing process. This return is the profit for the business. Without this information, correct decisions are not possible.

Consider a multi-plant organization in which the vast majority of the processes are similar. However, each site has handled asset-related data differently—some well and, all too often, some not so well.

This data at each site creates site-related asset information. The problem is that while the assets are virtually identical, the information provided where critical multi-plant decisions are made is not! How can senior management gather the correct information, make the needed analysis and resultant sound business decisions to drive asset optimization, and maximize return across the business?

The answer is that, in this scenario, they can't easily achieve this goal. At least they can't without a comprehensive effort to standardize the data and its associated information across the system. Proper data management at the outset could have eliminated this problem. As the CEO of this company, how would you feel knowing that asset optimization was needed, but difficult to achieve and completion highly unlikely over the short term? The worst case scenario here isn't that your company can't achieve this needed goal, but that your competition can achieve it.

The bad news is that lack of proper data management is all too prevalent across industry. The good news is that organizations are beginning to recognize the problem and address the situation. This book will help you learn how to address the issue of asset-related data properly so that the business does not suffer.

2.2   Who Are The Stakeholders?

If you were to ask, "Who are the stakeholders in the asset data management arena?" the initial answer would appear to be those working on site either at the middle management or the front line level. After all, they are the ones who use the data on a daily basis to run and maintain the plant properly. This is a narrow view of reality and actually quite far from the case! The importance of asset data and the related information it conveys is critical at all levels if the business wants to drive maximum return on its assets. Consider the asset data model seen in Figure 2-1.

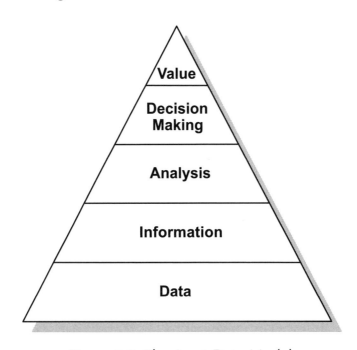

Figure 2-1 The Asset Data Model

Data—the raw granular basic elements that pertain to a specific piece of equipment—need to be correct, but alone has no direct business value. Data is analogous to the foundation of a building. By itself it does little, but when a structure is placed upon it, the criticality of having a sound foundation is readily apparent. Otherwise, the building will fall down.

The next level—information—is the amalgamation of data into something that is immediately important to the execution of the work. In fact, without asset-related information, work can't properly proceed, whether it is maintenance, operations, or any other associated function. The vital importance of good asset information at the workforce level is why, when asked about stakeholders, one immediately envisions this level of the organization.

Yet information isn't just needed to perform the actual work. Middle management requires accurate, usable, and fit-for-purpose information on a daily basis in order to conduct analysis at many varied levels. It is also their responsibility to prepare the analytical data in formats that enable senior leadership, all the way up to the CEO level, to make sound business decisions. These decisions, if made with the correct data, are the decisions that will drive maximum return on assets and ultimately profit for the business.

Not only are all levels of the organization stakeholders involved in the data integrity process, but so are the upstream suppliers and downstream customers. Without proper data delivered by the suppliers and delivered to the customers with the product supplied, data-related problems will be compounded due to possible inaccuracy or lack of completeness.

So the answer to the stakeholder question is really that asset-related data is important at all levels of the organization and across the supplier / customer interface. Without it, the whole system will never run better than sub-optimally.

2.3   Why We Wrote This Book

Although there are islands of data excellence across industry, the more prevalent asset-related data model is one of mismanagement, poor quality, limited accessibility, and ultimately the loss of business value. To further complicate the problem, there is a general lack of appreciation of the criticality of this data and the enormous business value that is lost. We wrote this book to address this problem and to convince everyone from the CEO to the workers on the front line of the value that proper asset-related data management can deliver. It is time that this level of understanding take place throughout the entire organizational hierarchy and across the supplier / customer interface.

However, this book intends to go far beyond a simple understanding of the problem. Simply understanding the issue doesn't deliver the potential value that exists. Understanding is just the beginning. There is a great deal more that we deliver in this book, including all of the steps that follow recognition of the problem and the desire to do something about it. What we deliver is the way to achieve business value from something that seems to be so insignificant as to be overlooked, yet is a key to ultimate business success and optimum return on asset value.

This book is more than a wake-up call to improve asset-related data management. It will take you beyond that high level of understanding the problem to the more detailed levels such as planning, executing, and sustaining a mastery of data management for maximum business value.

This is a "how to do it" book that won't simply change how you think about this subject. It will also change your organization's culture so that proper care of your asset-related data won't be viewed as additional work. Instead, it will be viewed as how work is normally conducted in your firm.

2.4   Who Will Benefit

Everyone!

As we will learn in great detail throughout this text, the integrity of our asset-related data and the information it provides has an impact on every level of the organization as well as others outside of the organizational boundaries—suppliers, customers, and those indirectly impacted outside of the supply chain. With a global understanding of the value of this data, everyone therefore will benefit.

Here is a brief description of the benefits from the proper management of physical asset data:

❏ The hourly workforce, particularly those involved in operating and maintaining the company's physical assets, will be provided information that will enable them to work more productively as well as effect repairs that result in more reliable and trouble free operation when the assets are returned to service.

❏ Supervisors and planners will be better able to support the hourly workforce with accurate and complete information about an asset-related activity, reducing errors and saving time.

❏ Engineering will be able to identify like assets across an enterprise, analyze, and compare various operating and maintenance strategies as well as the varying results of those strategies, and replicate what works well at all plants, in essence solving for the variation across the fleet of assets.

❏ Middle management will be able to acquire asset information without wasted effort and will be able to utilize this information for various decision making initiatives with confidence that it is accurate and reliable.

❏ The sales force will be better able to win new business and acquire market share from competitors. They will know with confidence the capability of the physical assets to deliver that added product to the marketplace in compliance with quality standards.

❏ Senior management will be able to make enterprise level, multi-site decisions that will directly support the optimization of asset value across the fleet.

❏ The CEO will realize improved return on assets and shareholder value because the entire asset management process will function in a much more efficient, effective, and predictable manner.

❏ The suppliers will be able to spend far less time addressing asset-related data issues such as searching for information lost at the site level.

❏ The customers will achieve a higher level of confidence in the quality of the product delivered, knowing that data within the plant is accurate and generating high-quality business decisions that affect the goods that they receive.

❏ Government agencies at the federal, state, and local levels, communities, and others will be able to maintain a high level of confidence that business decisions that have an effect on them are being made with the accurate and timely information.

2.5   What You Will Learn

This book is written to enable everyone to gain a full understanding of the vital importance of asset-related data and the information generated through its proper management and control. For too long, this topic has not received the attention required, nor have the processes been put into place to provide for its accuracy throughout the entire life cycle of the asset. This book was written to correct this problem. It will teach you the following concepts that can and should be applied at all levels of the organization:

❏ The value of this data and the need to properly maintain it not only at the plant level, but also across sites and at the corporate level

❏ How to initiate a data integrity process if one does not exist

❏ How to analyze your current process if one does exist; also, how to improve it

❏ How to sustain the process so that it becomes integrated into the culture of the organization

Learning and gaining a deeper understanding of data-related concepts will change your outlook regarding your plant assets and the information about them. This book will do far more than show you the road to success, it will also accompany you on your journey should you chose to take it.

2.6   Chapter Synopsis

Chapter 1 The Business Case for Data Integrity

The need for a dedicated and ongoing effort to maintain asset-related data is often overlooked due to the seemingly more pressing maintenance and reliability problems of the day. This shouldn't be the case, yet it is difficult to show the significant importance of this aspect of the work. This chapter introduces the topic of asset data integrity with a business case that is hard for anyone to ignore.

Part 1   Understanding the Importance of Asset Data Integrity

The first part of the book will provide you with basic information and understanding of the importance of asset data integrity. It sets the stage for the detailed information that follows in Parts 2 and 3. It also positions you to take advantage of what you will learn.

Chapter 2   Plant Asset Information—A Keystone for Success

If you have purchased this book, you have started on your journey towards improving the data associated with your plant assets. However, this is not a one-person excursion; it requires that everyone understand the serious need to improve this aspect of the work process. This chapter explains not only what you will learn, but also what you will need to impart to others. It is the beginning of a challenging but worthwhile journey.

Chapter 3   What is Data Integrity?

When the term data integrity or any other terms associated with this functional area is discussed, people usually have a general idea of what is meant. Yet there is neither common understanding nor clarity across business lines or even within departmental functions. This foundation is provided and it is needed in order to proceed.

Chapter 4   The Asset / Data Integrity Life Cycle

Every plant asset was new at some point in time. Every asset eventually reaches the end of life and is retired. Along with each of these assets comes the data and information that is part of the initial construction, on-going maintenance, and simply reference material used by those making asset-related decisions. Asset-related data has its own life cycle and its associated process, which crosses numerous departmental lines. Understanding the process enables the identification of problems and the resultant corrective action.

Chapter 5 Data Integrity at the Task Level

Very few if any organizations have the optimum data integrity process in place. In fact, in many organizations there is a need for a major improvement. We didn't get into the position of poor data integrity through poor performance by those within our organizations. The reasons for the current problems are far more complex. We need to understand what went wrong at the task level so we can learn from the errors of the past and not repeat them in the future.

Chapter 6 Internal Outcomes and Impacts

For every activity related to asset data integrity there is an outcome—what happens as a result of the specific activity and what impact it has on the business. It is critical that we understand this chain of events because we need to make sure that the outcomes and impacts are favorable to the organization. This chapter deals with internal outcomes and impacts.

Chapter 7 External Outcomes and Impacts

In addition to internal outcomes and impacts, there are a great many external ones as well. It is vitally important to recognize them because they address customers, suppliers, the government, and others. Negative outcomes and impacts in these areas can be costly in many ways and need to be avoided.

Chapter 8 Information Technology (IT) Problems and Solutions

With multiple departments associated with the data integrity effort, along with the fact that most of these organizations work in functional silos, there is a definite possibility that you will be confronted with multiple databases, many with redundant data for the same assets. Although some of this data is differentiated by functional requirements, much of the base data is the same. The problem is that the databases in which the data is stored are usually not integrated and often not even synchronized. The result: mistakes and other errors that can cause serious problems. This chapter describes efforts that are currently evolving to address these issues.

Part 2 Building a Sound Data Integrity Process

Part 2 builds on the basic information provided in Part 1. This is essentially a "how to do it" section which will provide you with the tools you need to create the high quality data integrity process that is needed for success.

Chapter 9 Building an Enterprise-Level Data Integrity Model

Whether your data integrity effort is a simple set of paper files or a highly-sophisticated integrated asset database, understanding what a good data integrity model looks like is important. For those where improvement over the existing state is recognized as being important, the path to achieving the level of data integrity you seek will be provided. For those who believe they have reached their goal, you will learn that there is probably more to do to make things even better.

Chapter 10 Building an Enterprise Level Inventory Catalog Data Integrity Model

Chapter 9 is focused on plant assets. Chapter 10 makes the same case for an improved data integrity model as it relates to inventory, an equally important aspect of the overall effort.

Chapter 11 Data Integrity Assessment

How do you know which areas within your data integrity process are working well and which are not, leaving room for improvement? The answer is to conduct a data integrity assessment. The purpose of this chapter is to explain the data assessment concept and the benefits that can be realized from an effort of this sort.

Chapter 12 Data Integrity Assessment—Equipment

The cornerstone of the data integrity effort is having accurate and available data about your plant assets. This chapter discusses the data integrity assessment process as it relates to plant assets and the inventory and parts that support repairs to these assets.

Chapter 13 Asset Data Clean-up and Repair

This chapter discusses how to clean up and repair the data bases that store our asset-related information. It will help you correct data that is missing, in error, or not synchronized, and many other problems that are associated with the lack of attention and poor treatment of asset data.

Part 3 Sustaining What You Have Created

It is not worth the effort of building a strong data integrity process if you don't plan to sustain it into the future. Part 3 describes how to sustain what you have created so that data integrity becomes ingrained within your organization's culture and an integral part of how you conduct your business.

Chapter 14 Data Governance

Asset data integrity appears to be everyone's job, yet no one's job. Given the level of importance and value that this data delivers, it is time to recognize that there is a new job within the plant—that of the Data Integrity Specialist. This new job role is described along with how to convince the senior leadership team of the need. It also describes how to put this new position in place within your organization.

Chapter 15 Sustainability

It is pointless to address the issues associated with data integrity if you don't take into consideration the sustainability of the effort. The work that is required to provide your company with a set of asset-related data that enables sound business decisions is not a project. Yes, it has a beginning. Unlike a project, however, it has no end. Without a sustainability effort to address this issue, all of the work you have done will be for naught.

Chapter 16 Data Integrity Is Serious Business

The work associated with data integrity is truly serious business. This chapter concludes the text with a brief summary and the advice to get started in order to improve your asset information and drive improved business decisions.

2.7 Let's Get Started

Up to this point we have introduced the topic of this book, but have really only scratched the surface. Data integrity and the management of asset-related data is a complex, multi-functional task. As we have tried to point out initially, if done poorly, the business can suffer, and you may not even know why. However, if managed correctly, there exists business value and profitability that has gone unrealized for a very long time. So let's get started!

Chapter 3

What is Data Integrity?

> A critical aspect about clarity is that it
> enables focus on what is important.
>
> *Data Integrity Work Team*

3.1 Defining the Terms

Many of the terms that are used to explain data integrity concepts
are used interchangeably. This use of different terms can cause problems when one is trying to provide focused information about data
integrity. The manner in which the terms are used can be somewhat
confusing and often misleading within the context of any discussion of
this subject. For this reason, it is appropriate that we take time to
define the three key terms that will be used in this book along with a
deeper explanation as to their meaning. Once this is accomplished,
we can link terms together and head toward a fuller understanding of
asset data integrity.

The first of our terms is *asset*. For the context of this text, an asset
is any physical entity within the plant boundaries. Assets can be

pumps, motors, vessels, tanks and the like, but also buildings, roadways, structures, fences, and things that people typically would not consider or think about when they talk about plant assets. However, we need to address all of these assets if we are going to provide the right combination of asset data to contribute accurate information for the managerial decision-making process.

The second term is *data*. Data is the elemental information about a plant asset. It is often confused with information and spoken about synonymously by those who discuss its integrity. Yet data and information are very different. Data are the raw elemental components that serve to explain an asset. Information is the grouping of data elements in such a manner that they can be effectively and efficiently used by the various levels of management for decision-making and action.

As an example of the difference between data and information, consider a storage tank. There are a great many pieces of data associated with a tank, including: height, diameter, capacity, location, function, material of construction and a great many more. Data such as height = 50′ and diameter = 25′ provide basic building blocks which, when combined, yield information about the tank. Individually these are just data points. When combined into an information set, a tank engineer can not only make decisions, but also develop action plans to assure that the tank is reliably able to perform its function.

The last of the terms is integrity. By definition, integrity is "a state of being entire or more complete." In our context, integrity is data that is related to our plant assets and is recognized as being genuine, complete, and trustworthy. Combining the terms we have discussed leads us to our definition of asset data integrity.

Asset Data Integrity: A collection of points or facts about a plant asset that can be combined to provide relevant information to those who require it in a form that is entire, complete, and trustworthy.

As we go forward, it is important that we understand that data is actually a subset of asset-related information. However, it is the data's accuracy that provides the usable reliable information for managerial decision-making.

3.2 Data Elements

What are the components of asset data—the various elements that, when combined, provide information to the organization in support of the work process and organizational decision-making? Many elements need to be considered if one is to truly have a comprehensive understanding of data as it relates to plant assets.

The different components of asset data include:

- **Physical,** This type of the data describes at a very high level the type of asset with which we are dealing. For example, physical data would explain that something is a pump, motor, tank, or any other physical asset within the plant boundary.
- **Dimensional.** Every asset has associated with it specific dimensions that portray to the user size, shape, and otherpertinent facts that dimensionally describe it within the physical universe. Dimensional data is a requirement of the business and decision-making process because it explains the details about the specific asset.

- **Technical.** Data elements that address the technical aspects of an asset provide the user with specific information that is often needed for repair and or replacement. Examples include drawings, material type, P&ID identification, and others. The vendor or supplier most often provides this information because they are the ones that have designed or built the asset in question. The problem arises once the data is received by the owner. Many firms have no work process or adequate computer system to address data storage in a manner that facilitates retrieval. Therefore, even if technical data is provided, its accessibility could be a problem. The mismanagement of this technical data often leads an organization to decision making that is less than optimal.

- **Reliability Focused.** These elements provide the user with data concerning the reliability associated with the asset. Examples of reliability-focused data are data points concerning preventive and predictive maintenance and asset specific trend data that can be employed to identify developing failures so that proactive action can be initiated.

- **Maintenance Related.** Asset specific data within this category applies to the work that is scheduled (or has been completed) on the asset. It provides data about the repairs specifically focusing on time and material work requirements.

- **Historical.** These data elements provide the user with insight into the history of the asset. It is often very important to have this information because historic data provides the user with insight into what has happened to the asset over its life cycle. Historical data allows for improved asset care and avoidance of past problems.

- **Material.** Data elements related to material are often critical in that they describe the metallurgy or other items used to create the asset.

With regard to plant assets, knowing the material of construction is very important because it enables the site to address reliability, safety, and environmental issues during repair and operation. Two other aspects of the material element are 1) the location where the material used in creating the asset originated, and 2) how the asset was built or assembled

- **Location.** The physical location of an asset is also a critical data point. Invariably, all plant assets have a physical location, yet different physical assets can occupy the same location over a period of time. The physical location describes the specific place within the manufacturing process in which the asset resides. The difference between a physical asset and its location is often confusing. For example, let us assume that a critical pump (call it P-1) is the primary charge pump for the manufacturing process. P-1 is its location in the process, not the actual asset sitting in that location. The asset should have its own unique identifier because over time the physical asset will reach the end of its life and get replaced by a new pump. When the asset moves or is replaced, the asset number goes with it; the location, P-1 remains.

  For assets that never move, this concept is not an issue. The asset number and location will always remain linked. Not so for assets that may move, such as motors and pumps. For this reason, it is critical to provide two sets of identifiers: one for the physical asset and one for the location.

- **Photographical.** Pictures related to plant assets are taken over a period of time. They provide very valuable data to users who may not necessarily be able to take time to go view the asset. Photographical data provides that view. In addition, photographical data provides insight into what was found during past disassembly and other work initiatives. The problem associated with photographical data is not with the data, but with the people who take the pictures. The result is often a database loaded with pictures that are not identified properly. All too often, the photos have limited or no value. Careful identification and screening of photos is important if this data element is truly to add any value.

- **Financial.** Data that fits this category provides the user the ability to relate specific asset expenses directly both to the asset in question and to other parts of the equipment hierarchy. As financial codes are assigned to an asset and its location, we can record and store expense information as it is acquired. It is important that our storage application have a way to maintain specific asset costs as well as location costs. For example, a specific location within the plant can exist for years. It is important to know the total cost associated with the location. This information provides insight into the total cost of maintaining that specific segment of the manufacturing process. It can lead to investigation and corrective action if the cost is excessive; it can also provide justification for capital upgrades. Acquiring cost data for each physical asset that has occupied the location is also important. Because physical assets may change in basic design, manufacture, or other aspects related to the asset, knowing the asset's cost will provide insight into its reliability as well as other useful information.

- **Hierarchical.** It important to have data about an asset and, separately, about the location. It is also important to know how the asset exists within the hierarchy of all equipment within the plant. When combined with financial data, this data provides insight into how larger amalgamations of assets exist in relation to one another. This data is often referred to as data about sub-processing units, production lines, or similar groupings.

- **System: Operational / Process.** Asset-related data is not only hierarchical, but also assets and their locations are integral parts of the manufacturing process (sometimes referred to as *operating systems*). Examples of these systems include the manufacturing process or sub-processes, utilities, steam, electrical, and water. These processes often exist within a specific sub-process or in areas such as utilities across the entire plant. Systems data is equally important so that asset repairs, reliability issues, process upgrades, or other improvements can be addressed more holistically. It makes little sense to execute a difficult, time-consuming repair on a pump without addressing the other components of the system. In that way, when the equipment is returned to service, the plant can have a high degree of confidence that the pump will operate optimally; so will the system of which it is a part.

- **Safety.** Data related to safety is critically important when any work is being done to the asset. Safety data provides insight to the users working in both planning and execution as to the safety precautions that need to be taken when working with this asset. This data is very important in order to provide for the safe execution of the work.

- **Environmental.** This data connects the asset with the specific environmental concerns that need to be taken into consideration when work is being performed. For example, equipment in acid or corrosive service needs special attention when work is being performed; environmental-related data provides this information. Although this element of data is sometimes overlooked, failure to take it into consideration could lead to serious negitive consequences.

3.3 Taxonomy and Why Is It Important?

Although it is important to address all of the different data elements, doing so is not the end of the story. There must be consistency for all of the data elements used within the asset data integrity framework. There are many reasons why a consistent approach is important. First and foremost, you must be able to find the data when it is needed. If you can't find what you are looking for, then there is no way to create information, conduct analysis, and ultimately make intelligent decisions. You find data by having in place a standard approach to the storage of the data while new assets are added or modifications made to existing asset data. In this manner, we have assurance of consistency within the data integrity environment.

Another important reason for consistency exists within multi-plant companies. In these situations, consistent data across sites for identical assets provides those using the data the ability to search multi-site databases. For example, suppose you work in a plant that has several sister sites. Today you are in need of a pump seal for a critical pump. The problem is that there is a very long lead time for delivery from the vendor. However, having consistent data across all of your plant sites allows you to quickly discover that one of them has the part you need on the shelf in their warehouse.

This consistent data structure is referred to as taxonomy and is defined as:

 Taxonomy: A comprehensive data structure that permits consistent classification of any person, place, thing, or idea managed by a system.

Taxonomy is a term that is not often understood clearly when it is used in the discussion of asset data integrity—it is not a commonly used term in the reliability and maintenance vocabulary. Essentially what taxonomy means for a single or multiple plant company is that the data structure for all of the assets across the system have a single methodology for classifying the data. If this common classification process can be implemented, then comparisons across multiple sites can easily be made using this standardized method for accessing the data and its related information.

Conversely, it is easy to see that an improper or non-existent structure could cause issues as the users attempt to access data either within the site or between sites. It is important that organizations such as Maintenance, Production, Safety, and others all have a consistent way for classifying and naming and referring to the assets within the plant.

3.4 What We Are Looking for in Good Data

There are many components of good data. We think the most important one is confidence that the data is correct. If the data retrieved is incorrect, so will be the information that is created and any subsequent analysis and decision making. Not only does the data need to be correct, but it also needs to be correct the vast majority of the time. We don't believe 100% accuracy is always possible—there are just too many handoffs and interfaces both human and systems in the overall process. Therefore, some small amount of error is bound to find its way into the data mix. However, more than a small amount is deadly. If a large amount of the data retrieved is incorrect, word will spread quickly, leaving a perception that all of the data is flawed. After all, perception is reality.

Many years ago, one of the authors worked on a project designed to merge separate maintenance computer systems at each of the company's plant sites into a single "best in class" application. As part of this project, all plant assets were "walked down"—a person visited the asset and recorded the pertinent data as well as extracting additional data from the plant's files. Then, this data was entered into the computer system. The problem with this effort was that no sustainability plan was put into place to maintain data integrity. Over the years, parts of the asset database deteriorated. Probably about 10 to 15% of the data was incorrect. After all, most of the plant's assets were stationary. They did not move nor were they replaced. The problem was that people did get incorrect data from the system. Many of them were very vocal about it because it created wrong information, analysis, and, at times, wrong and embarrassing decisions. The result: no one believed that any of the data was valid, thereby reducing the effectiveness and efficiency of the organization.

In addition, and in support of the concept of user confidence, other elements of good data need some explanation. These are called the Elements of Data.

- **Completeness.** All of the required data about an asset is available. Having only a part of the needed data will not serve the needs of the user. Instead, it can result in just as much productivity loss as if none of the data was available.
- **Conformity.** The data conforms to the existing taxonomy, making storage and retrieval a relatively simple task.
- **Consistency.** The approach to data handling is consistent. Everyone involved knows the process to store new data or modify what exists as it is altered. As we will learn later, providing a wide range of people with the ability to alter data is a dangerous strategy if consisten-

cy is to be maintained.

- **Accuracy.** This element is closely tied to the element of confidence. Accurate data is at the heart of the data integrity process; it promotes the use of what is available. Accuracy works on two levels and each is important. First is the accuracy of the asset's specific data. However, there is also a higher level, which addresses the accuracy of the asset's status. It is important that the user be able to check if the asset is in service, idle, or out of service. Often this piece of data is not well maintained, leading to confusion during work execution.

- **Duplication.** Many assets are duplicated within the various databases in use. It is important for these duplicates to be recognized within the data set. By eliminating them, it promotes inter-change-ability and optimization, not only for stocked parts, but also for the assets themselves. Several years ago, one of the authors worked in a plant that experienced a fire on a critical motor. Lead time was excessive. We were faced with a serious problem because of our inability to return the line to service. Fortunately, we had detailed data that indicated which assets were duplicates. As luck would have it, one of these was the one out of service. We were able to replace the motor and return the line to service in very short order. Without this element being part of our process, considerable profit would have been lost as we waited for the vendor to deliver a replacement.

- **Timeliness.** Having data available is only part of the equation. The data also needs to be available in a timely manner or the users will go elsewhere to obtain what they need. Timeliness of retrieval has a great deal to do with the system that is used as well as the taxonomy employed.

Consider how things worked in the "old days" before our sophisticated computer systems. Someone who wanted data about an asset
would fill out and submit a request; a clerk would process it.
Someone in the file room would then pull out the data, which
would then be presented to the requestor. In older days, this process
was considered to be timely data retrieval. In today's world, this
whole process has been reduced to a few clicks on your PC. Now
waiting a few minutes for data access on a slow system is unacceptable and violates the element of timeliness.

- **Maintainability.** A key element of data is the ability of the plant to
maintain it over time. This enables the site to adjust or change asset
data as changes are made or assets are replaced. Many companies
view this as an integral part of the job of every person who interacts
with the data. Infrequent use by these individuals will lead to degradation of the data and directly to problems of confidence. The element of maintainability is an important work process issue; it needs
a focal point for the effort and rigid processes for additions,
changes, and deletions.

- **Interfaced Data.** Few sites have fully integrated databases. For this
reason, the key identifying data elements need to be identical in
order to permit interfacing of the databases and their associated
data. Without this commonality built into the data elements, tools
like web portals and various forms of middleware will not be able
to extract data from multiple applications and present a unified data
set to the user. The second problem that arises without an interfaced
set of systems is having the data elements in the different applications—although having the same defined data element, you may
have different data residing in that element.

- **Fit for Purpose / Usability.** Asset-related data must be correct for the purpose it is to be used. If it is the wrong type of asset data, then the value to the user is compromised. For example, what value would machinery mechanics have for detailed trend analysis data when what they were seeking was information about reassembling the asset.

  The key is that all of these elements need to be addressed as a set of inter-dependent requirements for proper data integrity. If one or more of these elements is missing, then the likelihood is that the data being made available at the plant level will be suspect.

3.5 The Downside of Poor Data Integrity

It is a serious problem when the correct data is unavailable for the creation of information, proper analysis, and sound business decisions. Why? Breakdowns in acquisition, development, and use of data will ultimately result in lost business profit. There is another aspect of this problem that is equally serious. If people cannot readily acquire the data they seek, what do they do? In the case of asset-related data, they tend to save the data that they acquire during their various search efforts, knowing that it is correct and immediately available in their desk drawer or on the personal hard drive. Other people who are more familiar and skilled in development of personal databases in applications such as Access will take data storage one step further; they will build their own personal databases. Many of these people will even make their databases available for others perpetuating and expanding the problem.

There are several outcomes associated with these scenarios. First, the last thing we need is another database when our goal is to create a single data repository that is available to everyone. These additional databases create confusion; over time they become out of date and misleading. The reason is that the creators seldom maintain their database for an extended time. Eventually they get a new job, leaving their former job and database behind. The larger impact is twofold: 1) the database is not maintained and 2) there are users who believe it is accurate and will use the data in their decision-making process.

In a former job, one of the authors was directly exposed to this problem. An engineer had created a database of critical flanges (flanges that were part of dangerous piping configurations) and the specific technical and work processes required when working with these joints. The maintenance planners believed the database was still accurate. Unbeknownst to them, the engineer who had created it had moved to another job. He was no longer maintaining the database. Furthermore, the procedure for working safely with these joints had changed to a far more stringent process. Fortunately, for everyone, the problem was detected before work was executed. It is easy to imagine the potential damage and risk to the plant and its personnel had this use of the secondary database and its incorrect data not been uncovered.

3.6 A Word About Information Technology

Although the concept of data integrity directly addresses assets and their associated data, one cannot ignore Information Technology (IT). Unlike our past, where everything was stored as paper "hard copy," information technology has made data storage and retrieval easier and faster when applied properly. A good IT system allows us to address the Elements of Data as well as support controlled taxonomy. However, like any structure, a good system needs to be built using a strong foundation and well thought-out architecture in the construction. That is where the term data architecture was developed.

Data Architecture: A comprehensive data model that encompasses all the business processes and applications used to manage and run the enterprise. It is comprised of a detailed plan or design that maintains data integrity while permitting different uses across business functions and applications.

As you can see from the definition, data architecture applies to the entire business. Although many businesses have well-thought-out data architecture, many do not. This should not stop us from developing a data model and supporting architecture for asset-related data although the path will be far more difficult if a comprehensive plan does not exist. The main reason is that we may build something that works for our assets, only to have it become obsolete when the business finally develops the comprehensive data architecture.

But how do you assure yourself that the data architecture stays intact? As we have started to learn, there are many reasons the data architecture can be compromised and begin the process of deterioration. To avoid this, application functionality has been combined with work process. We are seeing a growing understanding of the importance of not allowing the data and its storage to suffer from neglect. This concept is called *Master Data Management.*

Master Data Management: Application functionality and business processes that govern or control the use of values in attribute fields. The purpose of Master Data Management is to ensure the integrity of the Data Architecture.

No matter what processes we assemble in order to create proper asset data integrity for a business, careful attention needs to be given to the information technology and the data architecture. It is not the intent of this book to discuss this topic in detail. Instead, the purpose of this book is to provide a data integrity model so you can create, build, and maintain a data integrity process. In addition, you need to be aware of the process and tie it closely to the existing data architecture and information technology, This is very important because the potential that IT can deliver is continually expanding.

There is another overriding term that is important in that it defines the intent of the entire data integrity process—it includes both the user and systems sides of the equation. The process by which this all comes together is referred to as *Enterprise Content Management* or ECM for short.

ECM is what this book is all about. Throughout the text, you will find the concepts and strategies to provide you with quality ECM within your organization.

 Enterprise Content Management (ECM): The strategies, methods, and tools used to capture and manage the documents related to organizational processes.

3.7 Understanding Data Is Just the Beginning

At this point, you are beginning to understand that maintaining asset-related data integrity is no small task. Most certainly, it is not one that can be handled on a casual basis if you want to be able to provide data; information, analysis, and value-added, decision-making capabilities based on data to the organization.

Chapter 4

The Asset / Data Integrity Life Cycle

Everything has a beginning, middle, and an end.
If you try to address each of these parts
individually, you are going to have trouble.

Data Integrity Manager's Forum

4.1 About Life Cycles

We are born, we age, and then we die. This process has been referred to as the life cycle of all living things. At different stages within our personal life cycle, we change, often in very drastic and notable ways. Compare a young child, a teenager, a young adult, and a senior citizen. The same person exists, but one who is very different at the various stages of their life cycle. As we age, natural changes take place. These changes manifest themselves in differences such as height, weight, and other attributes associated with our physical makeup.

Just as with human beings, the world of assets and asset-related data integrity is not static. It constantly changes as new assets are added to the physical plant and others are retired. To further complicate the process, data elements are continually changing even while the asset itself remains the same. Usually these changes are the result of repair or reliability efforts that are designed to improve the effectiveness and efficiency of the asset. Other changes that affect data integrity are corrections to erroneous data elements that are discovered. Operational changes that are implemented (such as pressure setting changes to relief valves), process changes that cause the re-rating of equipment, and other alterations to the data also occur during the life cycle of the asset. Last but not least, data is continually being added to further enhance the ability of the system to create information for analysis and, ultimately, business decisions. Examples of these additions include work order history, proactive and preventive maintenance results, and often the pictures that provide visual information about the asset.

Therefore, because we are dealing with dynamic data within a dynamic asset environment, it is important that we understand the process that addresses data integrity at all levels. After all, how can we expect our leadership to make sound asset-based decisions if the information that they receive is inaccurate and out-of-date due to our failure to keep up with the dynamic changes that impact the assets?

To provide the needed accuracy and timeliness of the data, we need to pay careful attention to the asset lifecycle and the data that is changed during the asset's life. An asset lifecycle is defined as follows:

Asset Lifecycle. The various stages that an asset passes through as it ages, from its initial design, specification, and purchase and commissioning all the way through to its removal from service and ultimate demolition.

4.2 The Asset Lifecycle

Initially an asset and its associated data are nothing more than an idea or concept in the mind of members of the management team. Often the individual assets are part of a larger initiative such as a new plant, a new process within a plant, or a major modification to an existing process. All of these various efforts employ new or redesigned assets to improve the business and drive additional value. However, much if not all this value can be lost either in the short- or long-term if proper treatment of the asset and its data are not addressed throughout its lifecycle. This begins before the asset even takes physical form.

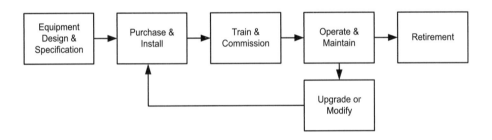

Figure 4-1 The Asset Lifecycle

Figure 4-1 represents an asset's lifecycle—one that closely parallels our own. Our understanding of the lifecycle of an asset will be even more important as we expand the discussion to that of the data lifecycle, which is closely tied to the lifecycle of its parent asset.

Equipment Design and Specification

Once the process design that is going to be created or modified is finalized, the next step for the engineers and designers is to determine what equipment is required—pumps, towers, drums, conveyors, and a whole host of other assets that enable the new design to function. However, this is only the first of the two steps necessary. Asset specifications also need to be developed so that the optimal equipment can be selected for the job.

Purchase and Install

After the equipment specifications are completed, they are submitted to the procurement side of the business to develop the bid packages and request bids from the potential vendors. This can take a considerable amount of time. However, at the conclusion of this process, the equipment is delivered to the construction site. The next step is installation. At this point, the asset is placed in its operating position within the plant. Depending on the complexity of the project, installation of assets following delivery can also take an extended period of time.

Train and Commission

This stage has two critical and parallel sub-stages. New or modified equipment requires training so that it is properly operated and maintained. Some of the equipment being installed could be rather complex and therefore require extensive and very detailed operator training. The training can be started even during installation, thereby shortening the overall elapsed time to get the operation running. There is also a maintenance component of training. Many assets are new, and similar pieces of equipment do not exist within the facility. In these cases, training of the mechanics who will service the equipment must take place; without this training, the proper repairs cannot be executed. It is also in this phase that the preventive and predictive techniques—which will be applied to assure asset reliability—are developed and included in the training program.

Once training is completed and the equipment installed, it then needs to be commissioned and placed into service. Although we would like to simply push a button and have the equipment startup, the real world is just not that simple. Commissioning requires a detailed start up and debugging process. This process can be difficult and time-consuming; at times, it requires modification to the equipment if it is to function properly.

Operate and Maintain

The next phase of an asset's lifecycle involves operating it as part of the process and maintaining it so the value it was designed to provide is indeed delivered. The "operate" part of this phase is really based on the training and the learning curve that the operators went through during commissioning. If handled correctly, failure due to poor or improper operation is unlikely. The other part of this lifecycle stage is maintenance of the assets. This involves the proactive plans developed and learned during training. It also involves the reactive repair processes that are occasionally needed. Although it is important to provide the correctly-designed equipment, the key to an asset's longevity resides in the Operate and Maintain phase. If operated and maintained correctly, an asset can have a very long life. The converse is also true.

Upgrade or Modify

Not every asset goes through its entire life exactly as it was originally designed and installed. Quite often, changes are made along the way. These changes may even be caused by manufacturer design improvements that necessitate modifications to the asset involved. At times, the asset is modified by the owner to improve its effectiveness and efficiency within the existing manufacturing process. Each type of modification or upgrade can not only affect the asset and how it operates, but also each type can have a major impact on the associated data.

Retirement

As with all things, there comes an end of life. No asset lasts forever. An asset eventually wears out and needs to be replaced. Sometimes its decline results in immediate demolition of the asset and removal from the plant. At other times, the asset is shut down and mothballed for future action. Another possibility is the removal of the asset's viable parts for use in other similar assets within the plant.

4.3 The Asset Data Life Cycle

Just as our assets have a lifecycle, so do the data that at a granular level define in detail the assets to which they pertain. Each of the blocks in the asset life cycle diagram shown in Figure 4-1 has a corresponding block in the asset data life cycle shown in Figure 4-2.

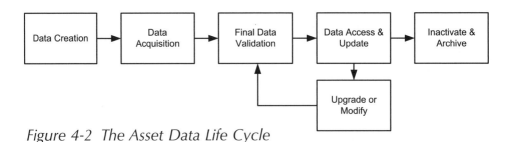

Figure 4-2 The Asset Data Life Cycle

The asset data life cycle is extremely important to our understanding of the concept of data integrity. It is also critical because this phase of the business, while often overlooked, has a constant need for attention to detail and accuracy. Without this understanding throughout the entire organization, the data suffers, the information provided becomes suspect, the analysis and decision-making become flawed, and serious business value can be lost. Even sadder is that without a full understanding of this life cycle and the important role it plays, many throughout your organization will not understand the root cause of the data integrity problem. Consequently, the organization won't be able to correct it.

Data Creation

As soon as equipment is designed, specific data related to that asset is created. It is a common fact that assets are not created in their final form the first time. Numerous iterations occur among the designers, engineers, suppliers, and others before the asset and its associated specifications are finalized and sent to procurement. At each of these iterations, data is created. It is critical that a process exists to capture this data in a controlled format so that, when the specifications are finalized, the correct data is included. Imagine the problems that could occur if this were not the case and the wrong data was provided to the asset manufacturer. Many assets have long lead times from specification to delivery. An important project could be seriously delayed were a critical asset built incorrectly as the result of bad data. It is even possible that a very profitable window of opportunity could be missed waiting for the replacement asset.

Data Acquisition

During every phase of a project or asset modification, asset-related data is provided to the owner. This usually takes place over the life of the project as the asset moves through its various design, fabrica-

tion, and installation phases. Often this data is provided electronically and, at times, on paper in a hard-copy format. Regardless of how it
is provided, the owner wants it stored in a manner that allows timely
access for reference, training, construction, and commissioning. This
transfer of data from the manufacturer and engineering company to
the construction contractor and, ultimately, the owner is a dynamic
process that continues up to and after final commissioning and
turnover. The interface between those supplying the data and those filing it becomes critical to the stages in the asset's life cycle that follow.
Depending on the size of the project and the number of assets
involved, the task of data acquisition and version control can be massive. If handled correctly, with the correct resources provided, acquisition of this intermediate data is a value-added effort for the business.

A project in which one of the authors was involved encountered
the downside of a poor acquisition and storage process. The value of
the project was in the hundreds of millions of dollars. The project
involved the addition of many new assets and the modification to a
great many more within an existing production facility. It was recognized at the outset that an asset-focused document management
process was needed. However, the process' complexity and the
required resources with specific skills to handle the incoming data was
not clearly understood. The result we encountered at the end of the
project was missing data, version changes that were not identified,
misfiled asset data, and large numbers of electronic files that were
unusable. Initially the problems associated with poor data acquisition
were not recognized. Later, after commissioning, the mistakes were
identified. Although small modifications to a single asset is not nearly
as complex, the need for an acquisition and version control process
still remains.

Final Data Validation

Some data has not changed from what was originally delivered during construction. However, some of it has changed during the construction phase. This data truly reflects the "as built" condition of the assets. If the owners have been careful during the project and maintained version control of the changing data, then they are ready to finalize the "as built" data received from the field.

The final data validation phase has two critical elements. The first is the maintenance of version control as the asset data changes during construction. Without this process in place, it will be difficult (if not impossible) to understand not only how the data has changed during the construction phase, but also which of the versions in the owner's possession is correct. The second critical element is the process to load the data into the asset records as finalized "as built" data while, at the same time, archiving the changes that occurred before the "as built" condition was finalized.

This final validation phase involves three critical parties. The first is the engineering firm. They're the ones that deliver the data, drawings, and specifications to the construction site. They are also the ones largely responsible for making the changes on the documents that are delivered during and at the end of the project. The second group is the construction contractor. This organization (often multiple organizations) makes the changes in the field during construction. They are also responsible for delivering the finalized "as built" marked up drawing at the end of the job. The third critical group is the owner. They have to receive the information and import it into their electronic asset data library.

It is critically important that these three groups work closely together and exchange data properly throughout the effort within a version controlled process. Without this close communication and a carefully-orchestrated process, it is entirely possible that the wrong data will be imported into the asset database as final.

It is also important to retain for a period of time, usually dictated by a company's legal department, all of the changes made prior to data finalization. In addition, all associated version control data—such as when the change was made, who requested it, and why—should be retained. Data saved in this matter can be of immense value if the asset experiences problems during its operation and the company wishes to conduct a root cause failure analysis. It is also often required for legal action to settle claims by those involved in the project—if a problem were to occur.

Finalized data is also necessary to upgrade or revise training manuals that include this data. After all, in order to assure proper operation and maintenance of the equipment, you want to train both operators and maintenance mechanics using the correct data and information it provides. Again, version control is important because the operators and the maintenance mechanics may have been trained using out-of-date data. This needs to be corrected to assure proper operation and reliability-focused repairs.

Data Access and Update

The data access and update phase is the longest in an asset's life because it extends from the commissioning up to and including retirement. During this phase, the operators extract data to run the equipment optimally. Mechanics extract data to maintain it properly. Of even greater importance, however, is the extraction of data to create information and subsequent analysis for the executive decision-making process.

If we have executed the data acquisition process properly in the earlier stages of the data's lifecycle, our problems should be minimal. However, each of the steps in the asset's lifecycle is handled by different people passing data across business interfaces. We're not perfect. Mistakes will be made, whether they are in paper or electronic form. It is at this point, when data errors or omissions are discovered, that a defined and well-understood process needs to be in place to make the corrections. This process can't be left to chance.

Many years ago, one of the authors was managing a work project that included the disassembly of a critical piece of processing equipment. When the work crew went to reassemble it, it was discovered that the gaskets in the asset's bill of material had been entered incorrectly. Therefore, the gaskets delivered to the field were wrong. As a result, we were delayed several days while new specialty gaskets were fabricated. At the conclusion of the job, we all agreed that the bill of material needed correction. Unfortunately, there was no process to assure corrections were made. Several years later, with a different crew reassembling the same asset, they encountered the same problem and the same painful schedule delay.

Incorrect data can often have serious consequences when it is converted into information for analysis and decision-making at the executive level. One particular project required shutting down for repairs a critical processing line; this outage was going to cost the company considerable lost profit. Nevertheless, the line had to be shut down because of equipment problems that were negatively impacting production. As you can imagine, a great deal of planning went into this project, all based on asset-related data extracted from the organization's asset database. To add to the problem, this line had been constructed only two years earlier. This outage was the line's first major one. Because the construction was so recent, the asset information should have been accurate.

The problem was that the plant did not do a good job of data acquisition, version control, and data finalization—all resulting in inaccurate data, faulty decisions made during planning, and a poorly executed outage. The data errors were not recognized until we were involved in the outage and then it was too late. Needless to say the outage was not optimal and a good deal of profit was needlessly lost. It would have been far less expensive to acquire and load that data correctly in the first place.

Upgrade and Modify

Changes to asset-related data do not just take place during construction. They also take place during operation and maintenance of the equipment in order to improve its operation or reliability. Yet the problem is the same. The asset data needs to be maintained in a timely manner as these changes are made to the equipment. Simply updating the data, however, is entirely insufficient. Depending on the changes that were made, there is a great deal of communications about the change that is required.

❏ Operations must be aware of the change because it may affect how they operate the equipment. The change also needs to be communicated to those involved in operator training so that, when necessary, the training manuals can be updated to reflect the new information.

❏ Maintenance also requires communication at several levels. The change could alter the maintenance procedures for repair. The change could also have an impact on planned work as well as proactive and predictive tasks associated with the asset.

Furthermore, communication needs to be made to the craftsmen who are directly or indirectly associated with the asset—these changes may affect how they maintain the equipment. Even crafts such as scaffold builders, insulators, painters, and others need to be included because changes to the asset often impact how these support group tasks are performed.

❑ Safety precautions are associated with almost every plant asset. As aspects of the asset's data change, so must the safety precautions for both operations and maintenance. Not communicating changes to those involved with safety issues could have serious consequences.

❑ The Environmental department is also impacted by asset changes. For example, these changes could result in modifications to environmental equipment checks, preventive maintenance, and legally-required compliance issues. Because failure to comply with environmental regulations can carry with it severe fines or other penalties, it is imperative that changes to the assets are addressed in a timely fashion within this arena.

❑ Engineering is always involved with equipment upgrades or modifications. Engineers use the data resident within the asset databases for design and eventual development of new assets, processes, and even new processing lines. This case is a classic one where data that is not updated can have serious financial implications on the business because the wrong information is used in the development of new designs, upgrades, or modifications.

❑ Materials Management is involved with the parts associated with the plant assets—when assets are altered, so are the parts that are used to maintain them. A great many of these parts are listed in the asset's bill of material. Our earlier example about gasket changes demonstrated a case where the parts requirements changed, but the bill of material was never updated. The result—the automatic ordering system that processes a stores requisition ordered the wrong material, with a major financial loss as a result.

Because of the vast list of required communications and system upgrades, a large majority of those in all industries have put in place a management of change process also known as an MOC (Management of Change). This process requires a sign off by all those affected by any change that has even the smallest impact on the manufacturing process. In most industries, the MOC process is a legal mandate and audited both internally or by OSHA (Occupational Safety and Health Administration). In the recent past, failure to establish and comply with the MOC process has led to major loss of equipment, serious injury, and even death in some cases.

Inactivate and Archive
Every asset reaches the end of its life cycle and is deactivated and possibly even demolished and physically removed from the plant. Equally important is the fact that the asset and its related data are deactivated. Most computerized maintenance management systems have built-in functionality to permit identification of both deactivation and demolition of an asset. The important issue is that data cannot simply be deleted and rendered unrecoverable, even if the asset itself is completely demolished.

- **Deactivation.** In this case, the asset has not been removed; it is usually just shut down and, at times, mothballed for possible future use. For this reason, the asset-related data must be retained for as long as the asset is in the deactivated state. The data may prove very useful if the asset is to be reused in another plant location, saving the need to purchase new equipment. Also, as the authors both have discovered on a personal level, these idle assets can bail a company out of an emergency situation where a duplicate asset needs to be replaced or the parts from the deactivated asset can be reused when the necessary parts are not readily available.

- **Demolition.** Even after the asset has been physically demolished, there is usually a corporate requirement that all of the asset's related data be retained. This retention period can be years. It is necessary to be able to access the data and be able to defend against possible lawsuits that include discovery of the asset's related data. For demolished assets, it is only after the corporate-mandated retention period that the data can and should be destroyed.

4.4   Why the Asset Data Lifecycle is Important

There is one overriding reason why it is very important to address both the asset and asset-related data lifecycles—they have a very close tie to business profit. Yes, it is true that data accuracy leads to more effective and efficient work, increased reliability, and improved safety and environmental performance. But these aspects of our business are directly linked to the profitability of the corporation. When we think of business profit, we immediately think of sales, profit margins, and all of the other financially-related components. Infrequently, if at all, does data integrity of our assets cross our collective minds when profitability is discussed. Yet the reality of the situation is that poor or inaccurate data leads to inaccurate information, faulty analysis, and possibly decisions that are certainly not contributory to business profit.

Therefore, attention to asset and asset data life cycles and also to the work and resources required to maintain the life cycles at all stages is so important. Responsibility for this attention should not be part of just the line function. It should also be shared throughout all the executive levels up to and including the CEO.

4.5 Roles and Responsibilities Within the Asset Lifecycle

One of the authors once led a project to install a state-of-the-industry computerized maintenance management system at a company where he worked. Along with the computer system and work process aspects of the effort, we also spent a considerable amount of time and energy validating the assets and the associated data. The time and effort that was spent involved a considerable number of resources, contractor support, and money. The day we went live was the day that the asset-related data was the most accurate; it was downhill from there. The reason was that there was no group or individual that had primary responsibility for data integrity. For new projects, completed after "go live," the engineers who acquired the data believed it was the responsibility of Maintenance to process the data. Of course, Maintenance believed it was the job of the engineers. As a result, over time additions, corrections, and changes of all sorts were never made.

The same was true for changes made during maintenance activities. In this case, everyone believed it was simply a clerical job and paid this task little attention. This line of reasoning—that it "isn't my job"—also appeared within all of the other departments that were exposed to the changes to the asset-related data. In turn, the plant's data degraded over time to the point where it lost all credibility and was no longer used.

Things don't have to follow the path of the example just cited. Awareness of the life cycles of assets and their associated data is a start. Once global awareness of the value of maintaining the data and the downside of not maintaining it is understood, there needs to be something done about it.

There is a solution to this problem. You need to make data integrity—whether entry of new asset data or upgrading existing data—a part of every project and every work initiative. This approach makes data integrity everyone's role. On the surface, this seems like a logical approach. However, it has major drawbacks if not carefully managed as you are expecting each group to independently execute their portion of the effort. These potential barriers include:

❏ Infrequent work in the data environment leads to erroneous entry, formatting, and inaccuracy. The problem simply is that not doing this type of work on a consistent basis results in loss of the basic skills of how to do the work correctly.

❏ Having a broader population of data handlers carries with it the fact that not everyone has the same level of commitment to the task. As a result, a task which should be handled in a standard fashion is not handled properly and often simply not done.

❏ A large number of individuals working with the data makes oversight difficult. More often than not, there is no audit trail. Therefore, if errors are discovered, there is usually no way of finding out who made the mistake so that corrective action can be taken.

❏ Inexperience working with the data could result in improper filing or even possibly data erasure, with no recognition of the problem until it was discovered much later. This problem is serious because erased or misfiled data is often not recoverable.

The problems associated with having data management be everyone's responsibility clearly has many issues. In fact, this lack of a single point of ownership can cause almost as many problems as not maintaining the data at all. The truth is that for optimum data integrity, the role and related responsibilities need a person or group of persons responsible and accountable for this phase of the business process.

For example, specialty engineers have a role during the equipment design and specification stage just as maintenance mechanics have a role at the operate / maintain level. However, there is a need for a focal point role that has a level of responsibility transcending all stages of the life cycle. In that way, each individual role can bring their expertise to the process while a single role is accountable to pull all of the pieces seamlessly together. This topic will be discussed in detail in Chapter 15 on asset governance.

4.6   It Is Never Too Soon To Start

Recognizing the importance of the asset and associated data lifecycles is something that most organizations understand at some level, and, in many cases, have done something about. To create and maintain a high degree of data integrity throughout the asset's life cycle is difficult. Most organizations fall into one of three categories:

1. You have a sound understanding and have implemented a process to address this critical part of the business.
2. You have an understanding of what is needed and have made an effort to address the process; however, improvement is still needed.
3. You're just recognizing the importance of this aspect of the business and are in a state of panic as you realize the vast amount of work and effort required.

It is never too late to increase your level of awareness and attention to this subject. Each of the three types of asset lifecycle awareness carries with it things you can do to maintain, improve, or initiate the data integrity process.

A Mature Process is in Place

In this case, there is no immediate danger. The lifecycle is understood and being handled at both the asset and asset-related data levels. The problem lurking on the horizon is one of sustainability, especially if your organization functions in a task-oriented reactive maintenance environment. Most likely, a mature data-driven organization has clearly defined roles and responsibilities and an overriding data caretaker responsible for the entire lifecycle. The problem is that this is a staff position. In times of organizational cutbacks, these positions end up being eliminated. This spells the long-term demise of the data integrity process. The sustainability of the effort is critical to long-term success. Senior management needs to be kept aware of the business value of this effort so that the process and those responsible for it can continue.

Improvement is Needed

This situation usually exists where a previously-sound process has degraded, and those with long-term responsibility to sustain the effort are no longer involved. As degradation continues, the organization will reach a state where data integrity doesn't exist and there is a lack of credibility even in what is available. This deterioration can't be allowed to happen; it leads to problems far worse than data that is simply not available. It leads to poor decision making and potentially lost profit. Those of us involved in the process need to show senior management how careful attention to the lifecycle is critical to the business' success. The value needs to be continually demonstrated so that asset data integrity doesn't get viewed as non-valued additional work and suffer the consequences.

Other organizations reach this point as they implement a program to address data integrity, but have yet to reach their established goals. In these cases, detailed initiatives and activities need to be established and their accomplishment set as a high priority for the business. Working toward evolving goals with management support is a clear path to a mature data integrity work process.

Initial Recognition Has Just Occurred

If you are just recognizing the need to address the asset and asset data lifecycles, you may be feeling overwhelmed. It is a massive task that most likely has been overlooked for a considerable amount of time. Just as every journey begins with a first step, yours begins with building a business case to convince the leadership team to provide sponsorship. Without leadership support, the effort is doomed. Many have been in this exact situation. Armed with a strong business case showing how data integrity was critical, and directly linked to business profitability, they have been able to acquire staffing and funding to get the effort underway. Then the realization sets in—support by the leadership team is only the beginning. Don't lose heart. The road to success will be difficult and take a considerable amount of time and dedicated effort, but it is worth it!

4.7 Life Cycle Links

In our discussion of asset data lifecycles, we reviewed the usual groups that one would typically think of when considering who would actually be involved at various stages of the lifecycle process. However, there are others that you might not consider.

Finance

In most systems, assets are linked to financial data that, behind the scenes, applies asset-related costs to various categories and subcategories within an organization's financial system. New assets need to be properly set up during the installation phase in order that these linkages be properly established and maintained. This set up allows for the capture of initial and subsequent costs during the operate and maintain phase of work on the asset. Asset-related costs are critically important at all levels of the decision-making process. These costs can provide information to decide whether to buy a replacement asset or spend money fixing the existing one. The wrong information can easily lead to the wrong decision and spending a great deal of money that doesn't need to be spent.

The financial links are also important as assets move. When asset relocation occurs, the financial codes tied to the asset need to be changed. This requirement, changing of the asset identification related to location, takes us back to an earlier discussion regarding asset location and the physical asset that resides in that location. Through careful maintenance of the financial links, especially during the maintenance portion of the life cycle, we can tell if reliability-associated costs are due to the location of the asset or due to the physical asset itself.

Suppliers

Those who supply the assets at the design and specification stage or during the asset's life cycle play a critical role. After all, it is the data that they provide which we use to populate our databases; the data ultimately become part of the decision-making process. If the data is inaccurate—not version controlled or simply not provided—our suppliers can indirectly have a major impact on businesses profitability.

Systems

The majority of organizations have some sort of system to manage asset-related data. Although technology is widespread in industry, some systems still have data stored in paper format. With that in mind, the Systems organization plays a critical role in the life cycle process. It is our ability to effectively and efficiently extract data in a timely manner that drives the information and decision-making process. After all, if you can't retrieve the data, it is the same as not having it available at all.

Customers

Regardless of what we produce, our customers are directly or indirectly part of the asset life cycle. They are the recipients of what we produce. If our data is wrong, there is a greater likelihood that what we deliver will be wrong as well. To look at this in a different light, we are the suppliers of someone else's asset and asset-related life cycle data.

4.8 Life Cycles as a Foundation

As we will learn in the following chapters, recognizing the importance of asset and asset-related life cycles and addressing them appropriately will set the stage for better decisions and, ultimately, increased profitability. The converse also applies. Lack of recognition of this important aspect of the business process will invariably affect profitability. As a result, you may never realize the underlying cause—a poor foundation of data integrity.

Chapter 5

Data Integrity at the Task Level

Task based change is only the tip of the iceberg.
Look deeper!

Data Integrity Work Team

5.1   Task vs. Strategic

For those of you who have a mature, sustainable data integrity process with close ties to the asset and asset data lifecycles, you are to be commended. You have in place the process and tools necessary to provide all levels within your organization timely and accurate information upon which to base sound value-added business decisions.

However, a mature data integrity process is not something to which many organizations can lay claim. Some never had it, some have recognized a business need and are trying to build it, and still others have had it and let it slip away.

Asset data integrity is a complex process existing at both the task and strategic business levels. Therefore, in order to create or restore a process such as this to the level required for support of the decision-making process, it is important to understand the underlying factors that make it work. This knowledge will allow you to recognize what you need to do to build a process where none exists or restore one that is failing.

5.2   The Data Integrity Transform

The data integrity process describes how data flows through a system of people who change its state at each step. Ultimately, at the back end of the process, the data emerges as information that feeds the decision making process. Often these changes of state are referred to as a series of transformations, or transforms for short, which include: data entry, the transformation process, and the end result where the data is organized and available for use. Figure 5-1 illustrates this data integrity transform process.

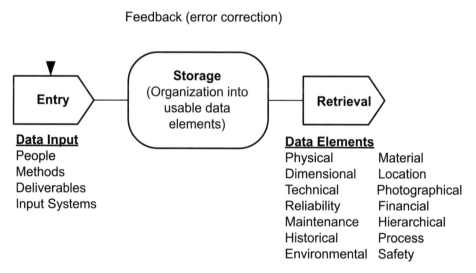

Figure 5-1 The Data Integrity Transform

In this model, data related to our plant assets is entered into a data storage system (usually one or more asset databases). It is then retained in a format that makes it retrievable by those who require the data to construct usable information for the business decision-making process. Retrieval of data when needed is important. The entry and storage steps, when handled properly, make retrieval possible. Let us review each step of the transform in some depth.

Entry

Entry is the manner in which data becomes part of the transform. As we learned from the asset data lifecycle, this data originates as new projects are constructed, and the data is delivered by the engineering and design firms that were contracted to complete the work. The other source of entry results from changes made to existing data or corrections to data that are found to be in error via the feedback loop.

Many problems can influence the entry portion of the transform. The important thing to understand is that if the entry step is flawed, then all the rest that follows will be flawed as well. The old adage— garbage in, garbage out—is very applicable here. The result of garbage out is far reaching as bad data yields bad information, analysis, and decisions. Simple problems with poor entry can be far-reaching and potentially damaging. Therefore, it is imperative that you understand the issues that come with poor entry. In this way, you will quickly be able to recognize them as they occur and do something to prevent the long-term problems that they can cause.

The problems associated with data entry fall into four areas: people, methods, deliverables (material), and systems.

- **People.** If people are assigned the task of data entry, they will most often do what is required of the job. The problem is that many organizations have not considered data entry to be a high priority task. As a result, no one is assigned to complete this task. It is left as a part-time effort spread across multiple organizations. In this scenario, many of those assigned will feel that it should be someone else's job. An additional problem is that some of those assigned also feel that it is a menial task and beneath them. In either case, the people assigned will not expend a great deal of effort and the entry part of the transform suffers.

A further people-related problem comes with working in a reactive work environment. In these cases, even though people may want to handle the data entry process correctly, they are so pressed with day-to-day work that the task of data entry is ignored.

- **Methods.** This area deals with the work process. Typically, a data integrity work process does not exist at the plant level. Without a clear work process, even if someone is assigned the task of data entry, there is no way to properly execute it. Without a clearly-defined process, people will do what they believe is correct. Their choices may not have anything to do with taking the necessary time to enter data correctly. This process does not just include entry of data, but also formatting it for future retrieval. If the data is not formatted correctly, it may get lost and not be accessible when needed.

- **Deliverables.** Another problem associated with entry is how it is delivered. Often a large project has vast amounts of data being delivered from the vendor. Unless the method and format of data delivery are specified early in the work effort, serious entry problems could exist. The data may not be delivered in the desired form (paper versus electronic), it may not be formatted correctly, it may not be what you want from the vendor, or it may not be version controlled. Therefore, you do not know if the data you receive is in its final form. It is critically important to specify the manner in which you want the data delivered. If not, the actual entry phase will be a serious problem.

- **Systems.** It is important that systems exist that will easily load the data. The preferred method is electronic delivery and the capability to automatically upload the data to the data storage system. This method eliminates many errors in transcription; however, it is not always possible. If you do not have the luxury of automatically uploading data, you need to be careful that the manual portion of the effort is carried out with extreme care. Even simple entry errors can cause major headaches when the data is needed. Systems are often built with front-end validation processes to reduce or eliminate these errors. Another less desirable way to handle error checking within your data system is with reports that will point out entry errors and allow for timely correction. However, creating reports as a means of identifying data is an after-the-fact solution and creates undesired rework.

Storage

Storage is the second stage of the data integrity transform. It refers to how the data that is entered is formatted and stored within the system designed as its repository. Storage is a very loose term that can range from a paper filing system up to and including a sophisticated data storage computer system. Regardless of the storage mechanism, the problems that need to be proactively addressed fall into similar categories.

- **Ease of Use.** The storage system needs to be easy to use. This includes entry and retrieval as well; without ease-of-use, people will create their own personal systems that they feel provide this capability. The problem here is that these personal systems are just that—personal. They take the form of personal databases all the way to storing data in one's desk drawer or a filing cabinet. In every case, the data is available to the person, but seldom to the organization. Not only is it not available, but its use can be questionable. Their database may be wrong because changes are made in the data, but not updated.

- **Standard Format.** The storage system also requires a standard format for all assets and asset-related data. Once stored, the data can then be accessed and retrieved. Wherever possible, this format—which is typically addressed during entry—needs to be standardized and, if possible, made unalterable by those entering the data. If this is not the case, people are left to their own interpretation of how data is to be entered. Proper formatting and successful retrieval will prove to be a difficult task.

- **Single System.** Wherever possible, you need to make every attempt to have a single system or one in which the data is integrated. A single system leaves you assured that the data you retrieve will come from a single source and from the source into which it was entered. An integrated system—one with different modules that are electronically linked—will leave you with a degree of confidence that the data entered into one module will be automatically updated within the other integrated parts of the application.

The danger arises when asset and asset-related data are stored in multiple non-integrated databases. In these cases, you do not know if an identical data element will be the same or, if not, which one is correct. This can cause serious problems if the data is different and the wrong data is used for analysis and decision-making. Another associated problem if they are different is how you know which one is correct and which one needs to be corrected.

One past client had three relief valve applications, each managed differently by three different organizations. They were the maintenance database, the database maintained by the inspection department, and the one managed by the process engineers. Each had data that was unique to the respective function. However, in many cases, even the core data (asset #, name, location, pressure settings, etc.) was different. The dilemma: Which one was correct and how could we fix the problem of synchronization of the common data? The solution actually involved extensive work to walk down each asset and make the necessary corrections. It took years to complete the task which, if we had provided standard formatting at the outset, would never have been necessary.

Retrieval

Retrieval is the final stage of the transform. When people seek asset-related data, it is usually needed promptly. Once acquired, the data needs to carry with it a high level of credibility. If these are not traits associated with the data integrity transform, people will figure out other ways to acquire the data—usually via personal databases or a desk filing system, both of which are as risky as retrieving data from a system with questionable accuracy. A good retrieval process is associated with the system used to store the data. Yet even a well-managed paper system will work well if the entry and storage parts of the transformer work properly.

Feedback

This portion of the transform process addresses data that is

retrieved and is identified as being incorrect. Without a properly functioning feedback loop to make corrections to erroneous data, it will
never be corrected. Again, without people designated as being responsible to address the corrections, the errors will never be fixed.
Although it is not acceptable to retrieve erroneous data, make business
decisions, and then suffer the consequences of suboptimal results,
how serious is it to a business to do this repeatedly? For the transform
to function properly, people need to have faith that the feedback
process works and that the errors they identify will be fixed. If not,
they will cease to identify them. Over time, the data will degrade to
the point that it isn't used because no one believes its validity.

5.3 Data Integrity Tasks

The failure of the data integrity transform is essentially the failure
to perform well within the specific stages: entry, storage, retrieval and
feedback. We have pointed out how some of these failures manifest
themselves, ultimately destroying the entire process and often the
organization's ability to make value-added decisions. However, it is
expected that an organization in a data integrity rut, observing these
manifestations of poor data integrity, would want to fix the problem.
How can these failures of the tasks that make up the transform be corrected? What needs to be done so that an organization can provide
those who need data with timely, credible data to feed the information
and analysis process?

Many firms think that they can address and correct task-related
problems by fixing the task—they are missing and not addressing the
root cause. Data integrity problems go beyond the failure to perform a
data-related aspect of the transform. They go all the way to what we

call the organization's frame of reference related to data integrity and other aspects of the work.

Although it applies to data integrity, the organizational frame of reference—how they focus on and conduct work—also applies to every other phase of the business. There are two essential frames of reference describing how organizations think about work:

- **The Reactive Frame of Reference.** In this mode, everything is an emergency or rush; requiring people to drop whatever they are doing and quickly address the emergency of the day. A good example of this perspective is reactive or emergency-based maintenance.
- **The Proactive Frame of Reference.** In this mode, things are well planned out and executed. Emergencies are rare. When they do occur, they really are emergencies and not just a group's way to get prompt action at the expense of the planned work.

This concept or frame of reference is important at the task level because it explains why the data integrity process works properly or fails to function as expected.

5.4   Reactive Data Integrity

The reactive data integrity frame of reference and how it functions to undermine an effective process are shown in Figure 5-2.

This diagram is often referred to as a reinforcing loop because it shows the steps involved in data acquisition, what happens when the erroneous data is provided, and what an organization operating in a reactive frame of reference does about it.

In Block #1, the organization requires asset-related data for a variety of reasons. Because the data integrity transform is not functioning as it should, there is often a protracted amount of time required (Block

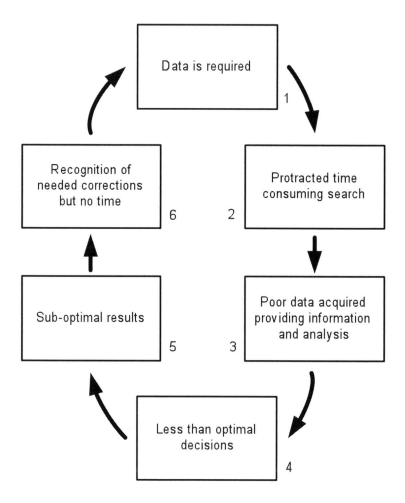

Figure 5-2 The Reactive Data Integrity Process

#2) to obtain the data. Not only is this frustrating to those involved, but it also has an enormous cost associated with the lost time. Once obtained (Block #3), it is often of poor or even unusable quality. Ultimately data that is unusable from the company databases will cause users to create their own credible (but not widely deployed) asset storage process and tools. However, the more significant problem is that poor data forces analysis with incorrect information or, as an alternative, best guess estimates. The results (Block #4) are less than optimal decisions and the associated result (Block #5) suboptimal results. Depending on what type of decision is involved, the cost to the business and associated lost profit can be enormous.

Of course, those at various levels of management recognize the problem. After all, who wants to be caught up in a process that delivers poor results? The challenge is not that people do not recognize the problem. They just do not know how to fix it. The solution is in the organization's reactive frame of reference. As indicated in Block #6, they understand the need, but there is no time to fix it as they struggle to overcome daily emergency issues. In a sense, the integrity of the data and the need to correct errors does not become a priority and, inadvertently, the loop is closed—the poor data integrity process goes unaddressed and lack of data integrity performance is reinforced.

This mode of operation should be unacceptable if you wish to have a high degree of data integrity in your plant. However, fixing it is not as simple as just saying, "From here on out, we will develop, implement, and maintain a properly-functioning data integrity process." The reason is that, to accomplish this goal, all of the associated challenges with the data integrity transform must be fixed— improved data entry, proper storage, timely retrieval of credible data, and a healthy feedback loop. By completing all of these tasks to correct the overall process, the organizational mindset will be altered, providing the opportunity to change the reactive data integrity process to a proactive one. Without engaging in this change process, simply mandating better data integrity will be a useless exercise.

5.5 Proactive Data Integrity

The change to a proactive mindset brings into play an entirely different reinforcing loop, as shown in Figure 5–3.

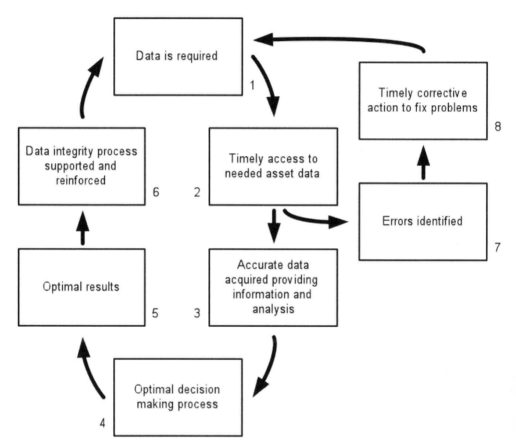

Figure 5-3 The Proactive Data Integrity Process

In Block #1 of this loop, data is required. Unlike the reactive loop, the data integrity transform is highly functional and delivers the data required (Block #2) in a timely manner. The data is then utilized to create information for analysis (Block #3). However, there is another information pathway, as shown by Blocks #7 and #8. With vast amounts of asset-related data in the system, the data extracted can never be perfect 100% of the time. The correction of identified problems and errors is addressed by the corrective loop of Bocks #7 and #8. In this organizational frame of reference, accuracy is important because the organization recognizes how it affects business decisions and ultimately profit. Therefore, when errors or omissions are discovered, timely corrective action is taken to correct the problem so that it will not happen again.

The data in Block #3 is converted to information and analysis, which then lead to optimal decisions (Block #4) and optimal business results (Block #5). Those at the highest level of the organization recognize that the decisions they are making—decisions that could affect the very survival of the business—are based on accurate and timely asset-related data. Consequently, they support and reinforce this process, closing the loop (Block #6).

5.6   From Reactive to Proactive

It is probably obvious to all involved that when they acquire poor data and make their decisions based on that data, something needs to be done to fix this entirely unacceptable mode of operation. For those operating in the reactive mindset with a nonfunctional data transform and process that do not promote data accuracy, it is not too difficult to agree that things need to change. However, change is not easy to achieve. Otherwise, everyone would do it as soon as they recognized and understood their problem.

Change needs to take place much deeper in the functioning of the organization. It is not as simple as deciding to change how tasks will be performed. Yes, it is possible for managers to mandate changes in how a task such as data integrity will be handled. But in that case, change will not stick! In fact, the organization may revert to the former process—the reactive mindset—as soon as the managers refocus their attention in another area of the business.

Still, it is possible to change. However, if change is made only at the task level, it will not work. What is needed is change at the strategic and cultural levels of the business model. If attacked at these two levels, the change will stick.

Chapter 6

Internal Outcomes and Impacts

It is important to consider outcomes and impacts; otherwise, you may do something you believe to be beneficial only to realize too late it isn't.

Data Integrity Manager's Forum

6.1 Indirect Impacts

The preceding chapters addressed the asset and asset data integrity life cycles. They described in some detail how asset data and its use were tightly interwoven with the primary organizations within a plant, such as Production, Maintenance, and Engineering. We also briefly discussed the connectivity that asset data had to other departments that, at first blush, might not even be considered as a part of the asset-related data framework.

This chapter is going to drill deeper into the examination of the other departments that not only provide input to the asset data integrity process but also draw data required from it to effectively and efficiently perform their jobs. Although these departments make data-related business decisions, in most cases, their decisions are indirectly related to the profitability of the organization. Nevertheless, these decisions are important. If made incorrectly, they may certainly have a negative impact on the business.

What does "indirectly related" mean to asset data integrity? To answer this question clearly, we must differentiate between those departments that have a direct relationship to profitability and those that are indirect.

Departments that have a direct impact include Production, Maintenance, and Engineering for the following reasons:

❏ Engineering develops new processes or upgrades those that already exist. Usually these projects are the result of detailed analysis and, when implemented, drive profitability.

❏ Maintenance's focus is on repair, proactive or reactive, which in turn keeps the equipment operating, which directly supports company profit.

❏ Production operates the plant processes that transform raw materials into finished goods, which is the underlying element of the firm's revenue stream.

The use of asset-related data within these departments is clearly linked to profit; erroneous data can easily lead to incorrect decisions and lost or un-optimized production.

The departments that indirectly apply asset-related data fit into the picture in a different manner. The decisions that are determined by these departments are somewhat disconnected from profit. These decisions directly support the workings of the department in question rather than create profit. They serve to avoid unnecessary cost to the business. The decisions do not add additional dollars to the bottom line; rather, they avoid loss.

Some examples are in order to make the point.

Safety
Many asset databases have embedded in them safety procedures and processes that are often extracted for work execution. What if the wrong information was entered into the system, extracted, and then used? The results could be disastrous.

Environmental
There are many activities within a processing plant that require environmental data be input to the asset database. Examples include preventive and predictive maintenance schedules for environmental equipment, alerts and alarm data telling production when they have an emerging environmental problem, and data from the various environmental tests that are either company or legally mandated. This information is also extracted for use in the decision-making process as well as for legally-required environmental reporting. Once again, incorrect data could result in environmental problems with associated legal costs and fines.

One could argue that these examples—and still other examples pertaining to other departments that have an indirect impact associated with data entry or retrieval—are all directly tied to profitability. It is true that examples can be cited that one would consider at least in a gray area. Therefore, what we need is a way to differentiate direct and indirect effects on profitability. To accomplish this, we need to ask a simple question related to the asset data used by a department within the plant. The question is: Does the asset-related data, if incorrect or misapplied directly, impact the ability of the plant to produce product and thereby generate a profit? A "no" answer leads one to the conclusion that the impact is indirect.

If you recall the data integrity transform from prior chapters, the concept of direct and indirect impacts allows us to expand it and provide a visual view of this concept, showing both direct and indirect input and output (Figure 6-1).

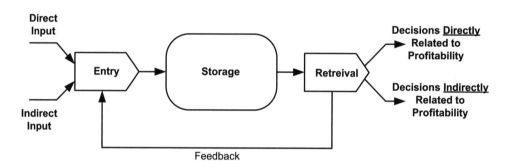

Figure 6-1 An Expanded View of the Data Integrity Transform

6.2   Decisions Are Just the Beginning

We have spoken about data-driven decisions and how they affect the business. These decisions are quite often made from the asset-related data retrieved by those who need it. However, there is more to the discussion because the result of any decision is some action and has results that touch both external and internal groups.

Decisions spawn outcomes. In turn, outcomes have impacts on those directly or indirectly involved in the manufacturing process. Both outcomes and impacts are important and play key roles in the data retrieval process.

❏ **Outcomes** are the results of decisions. They are what take place after a decision is made. Outcomes can be major, resulting in large-scale initiatives being developed which often can place the very survival of the business at risk. They can also be minor, focusing on specific work tasks related to the conduct of the business. Nevertheless, every decision has one or more outcomes.

❑ **Impacts** are how the various components of the business are affected. Outcomes create impacts for the company, suppliers, customers, employees, and virtually anyone or anything associated with the decision. For this reason, whenever possible, positive impacts need to be sought as the desired result of the outcomes from decisions that are made within the firm. Although this isn't always possible, it should be the expressed goal of the effort.

The concept of outcomes and impacts allows us to further expand the data integrity transform by understanding that both direct and indirect decisions carry with them these components (Figure 6-2).

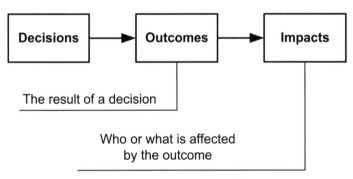

Figure 6-2 Outcomes and Impact

To clarify the chain of events that results from a data driven decision, let us consider an example.

Situation. A pump within the plant is experiencing a high rate of failure. The data retrieved from the asset database related to this asset and the root cause analysis of the failures clearly point to the mechanical seal and its current design.

❑ **Decision.** Based on analysis of the data, the decision is made to review the seal design and determine if there is a better seal available on the market to handle the current pump service and reduce the rate of failure.

❑ **Outcome.** An alternative design is selected from a different manufacturer. This new design appears to be better suited for the existing service.

❑ **Impacts.** Usually there is more than one impact and often not all of them are positive.

- Reliability improves because the new seal is properly designed for the application.
- A new training program is developed for the mechanics so that they will be able to install the new seals properly.
- Modifications are required to the pump's asset bill of material, reflecting the new parts.
- There is a cost impact on two fronts. First, the new seal is five times more expensive and this expenditure was not budgeted. Second, there is a significant saving due to reduced frequency of repair that more than offsets the cost of the seal.
- The new vendor experiences an increase in revenue from sales. Additionally, they now have the attention of the plant's machinery group and have convinced them to explore other uses of their products.
- The former vendor experiences a loss in sales as well as credibility because they should have recognized the problem with the design of their seal in the service to which it was applied.
- There is also an impact on all of the databases that contain information related to the new seal design and method of installation. The impact here is that these databases need to be

changed. Otherwise, incorrect data will be retrieved the next time work is conducted relative to the pump seal installation.

It should be apparent now that simply retrieving asset-related data is not the end of the story. It is what happens as a result of the retrieving of the data that is really important.

6.3   Indirect Inputs

An asset database is a dynamic tool with data constantly flowing into and out of the storage area. Earlier we discussed Production,

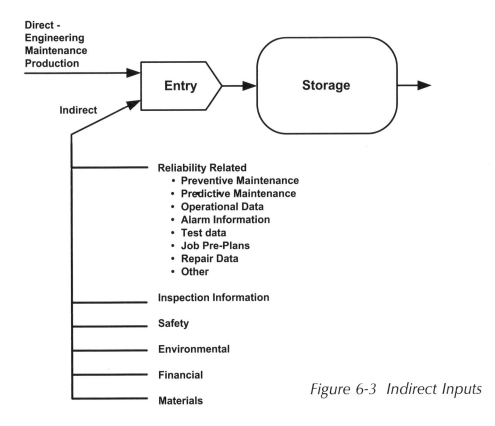

Figure 6-3 Indirect Inputs

Table 6-1

Category	Data Integrity Input
Preventive / Predictive Maintenance	· Work schedules · Procedures · Guidelines · Safety precautions · Historical · Future repair requirements · Estimates · Work details
Operational Data	· Process related information · MSDS (Material Safety Data Sheets) · Procedures · Guidelines
Alarm Information	· Settings –alerts / alarms · Protected equipment information · PM / PdM schedules · MOC information
Test Data	· Work history · Future repair requirements · Reports
Job Pre-Plans	· Work plans and estimates · Resource requirements · Safety issues
Repair Data	· Historical information · Reports · Trend data · Failure analysis (RCFA) · Future requirements
Inspection Information	· Historical data · Future repair requirements · Corrosion information · Asset history and end of life data · Reports · Pictures
Safety	· Procedures and guidelines · Audit information · Event history
Financial	· Asset codes · Depreciation information

	· Tax data
	· Cost analysis (asset related)
Materials	· Stock information
	· Direct requisition data
	· Purchase order history
	· Parts lists (stocking information)
	· Materials – historical data
Environmental	· Operating permits
	· Test results
	· Compliance schedules
	· Legally required reports
	· Penalty / fine information

Maintenance, and Engineering—those organizations that have a direct effect on the database. However, as depicted in Figure 6-3, there is also a wide variety of indirect inputs.

Indirect inputs are all internal. They come largely from departments and work groups that input asset-related data to improve how work will be executed in the future. These inputs also have a secondary benefit because they provide improved supporting data for asset-related business decisions. This data comes from feedback as a result of executed decisions or entirely new data as assets are added or changed.

Let is examine some of the indirect inputs so we can be clear on what is included in this rather broad category (Table 6-1).

At some point in time, indirect inputs will undoubtedly be used as outputs for the creation of informational packets to drive or, at minimum, support business decisions. Therefore, they must be accurate, timely, and easily retrievable.

From the list shown in Table 6-1, on page 100 it should be obvious that when we discuss asset-related data, the amount of input that can flow into the transform can be immense. Some of this data flows in automatically by virtue of how the computerized maintenance management system (CMMS) is configured. However, a great deal more is manually entered and not always into the CMMS. Many companies do not have a single system that handles all of the asset-related data. I have seen numerous cases where each department has an asset database that they use; quite often these are not synchronized with the CMMS.

This fact creates a problem for those who desire to input asset-related data for future use in the decision making process. Where does the data get entered? If entered in one database, is it also entered in all the others so that retrieval from all locations will yield the same result? Unfortunately for many plants, a single data integrity system or synchronized data entry to multiple databases is not the norm. It would be easy to recommend that those of you who have this problem simply consolidate everything into a single CMMS-like database. The issue is that this effort is neither simple nor inexpensive; furthermore, it is extremely time consuming. It is far better and much more practical to consider developing a strict work process that forces synchronization of the data so that it is consistent across all of your databases. The concept of a single repository for asset-related data, and how that data source should be synchronized to all consumers of the data, will be discussed in more detail in later chapters.

There is another issue on the entry side of the transform. Depending upon whom you ask, there are various opinions of how much data should be entered. Answers run the gamut from everything to only what is absolutely needed. However, the real answer is not a simple one; it will be addressed further in Chapter 15 Data Governance.

The problem goes beyond how much should be input. Other factors are at play:

❏ The organization's philosophy of what should be stored can change, causing large swings in the scope of the data integrity effort and "data holes" where data that is now required was not required in the past.

❏ The support of this administrative process can change as the leadership changes, providing more or less resources for the work.

❏ The sponsors of the effort change over time as organizations alter their structure. This change can cause problems because direction, focus, and overall guidance can change drastically.

❏ Legally-mandated requirements may be imposed on the company dictating the data storage requirements.

With all of these factors involved in what flows into the data integrity transform, you need to be careful that a level of consistency is maintained. After all, the data will eventually be used in the decision making process; we want the decisions made to be the correct.

6.4   Indirect Outputs

Indirect outputs from the data integrity transform, the decisions that they affect, and the outcomes and resultant impacts fall into five distinct categories (Figure 6-4) on the following page.

Data is retrieved because a decision needs to be made regarding the operating of the business. Every one of these decisions has outcomes and resultant impacts. There are three major categories: those that affect the business, those that affect compliance, and last but certainly not least those that impact the media. Each of these categories is different. The business and compliance categories each have subcat-

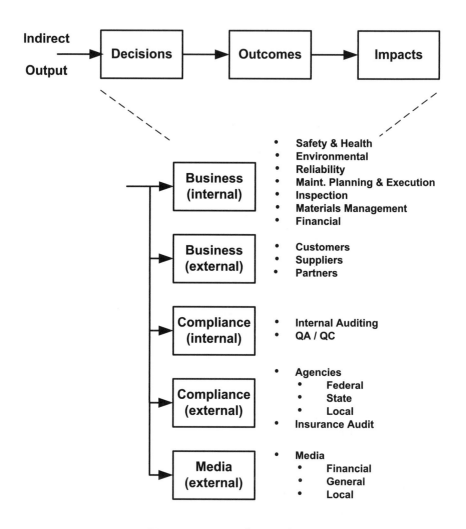

Figure 6-4 Indirect Outputs

egories focusing on internal and external impacts. This chapter will address the internal aspects; Chapter 7 will address the external ones including those related to the media.

Business—Internal

The data integrity outputs affect many internal groups that are indirectly tied to the data integrity process. These groups depend heavily on asset-related data to perform functions supporting their portion of the business. It is asset-related data that is converted into the information they utilize in their respective decision making processes. Good data yields good outcomes, many of which are heavily dependent on the input side of the transform. It is the bad data that ultimately causes problems, not only for plant reliability, but also in all of the areas shown in Figure 6-4. It is for this reason that good output is critical.

SAFETY / HEALTH

The asset-related data extracted are often the very procedures and guidelines that were input to the database. The reason for this is that these documents need periodic review to validate that they are still accurate. Quite often the practices, which change as a result of experience working with the asset, need to be modified. They often change as new rules and regulations are put into effect. For example, take the procedure for handling asbestos. Years ago it was quite different from what it is today. Safety data is also extracted to provide statistical and trend information related to specific assets. This data is then used to make decisions on whether or not to modify the guidelines and procedures in order to make the work environment safer for the employees.

ENVIRONMENTAL

A great deal of environmental data is extracted to provide information internally as well as externally. This data is used to show that the environmental assets are performing as required. Many of these assets operate under strict guidelines in order to ensure environmental compliance. When the data reveals that the process is out of compliance, prompt action is required to correct the deficiency. If these issues are not addressed, the company is likely to be exposed to environmental compliance fines, some of which can be quite substantial.

Asset-related environmental data also provides preventive maintenance schedules. Because it is critical to have environment assets performing at 100% all of the time, a great deal of PM information is stored and triggered on a scheduled basis to ensure that the equipment functions properly.

RELIABILITY

This asset-related data is usually tied to keeping the equipment functioning in an optimum manner. Typical output is in the form of trend data so that maintenance corrective action can take place prior to a major failure of the asset. Additional output is often tied to future work that was discovered the last time that the equipment was repaired. In these instances, work that could not be completed at the time of repair is stored and addressed the next time the equipment is taken out of service for repair. In this manner, it can not be forgotten and lost.

MAINTENANCE WORK EXECUTION

Asset-related work execution data takes the form of job plans saved from prior jobs. These plans include work schedules, resource and equipment requirements, work step timing, materials required, safety issues, and all of the other critical aspects required to perform the work effectively and efficiently. This data is valuable because it saves time and money for repetitive jobs.

Another often-overlooked piece of data shows how the asset in question ties into the larger operating process. All too often we make repairs to a piece of equipment, not taking the time to examine needed repairs to other assets that are part of the manufacturing process being taken out of service. Having data that shows how an asset is part of a larger system, and knowing what other work can be performed when the asset is out of service, has immense value to the maintenance organization.

INSPECTION

The Inspection Group typically concerns itself with fixed equipment and piping. The other departments—such as machinery, instrument, and electrical—usually handle their own respective classes of assets. Data retrieved by Inspection often deals with deterioration of the metallurgy of the fixed assets. This information is all historical, but it provides data that enables this group to make decisions on replacement or repair based on the trends that they observe.

MATERIALS MANAGEMENT

Every asset is composed of parts. A great many of these are part of the warehouse inventory; referred to as critical spares. The reason for this descriptive phrase is that they are critical to maintain the plant's operation and can not be acquired from the vendor as rapidly as they would be needed in an emergency. The shelf value of these parts can be considerable. There are also other parts that are utilized in the repair of the asset that fall into the category of general stock—materials kept on hand to support the repair of the asset. The information retrieved by the Materials Management organization is in the form of inventory levels and part usage history. This information allows them (working with the reliability organization) to optimize the value of the asset-based material that is kept on the shelf.

FINANCE

The Finance organization uses asset-related data for cost analysis as well as in the support of business decisions related to the value of repair vs. the cost of replacement. Quite often this information—historical asset cost data—can be used to show that there are significant savings associated with purchasing a new asset as opposed to the continual repair of an existing one. The finance organization also has to accurately monitor and report the value of assets as they depreciate.

Compliance—Internal

There are other areas upon which asset data has an impact. In most businesses that produce goods and services, there is a high degree of regulation (Chapter 7). However, smart companies are not satisfied with allowing outsiders to discover areas where there may be compliance issues. In order to uncover these issues so that they can be

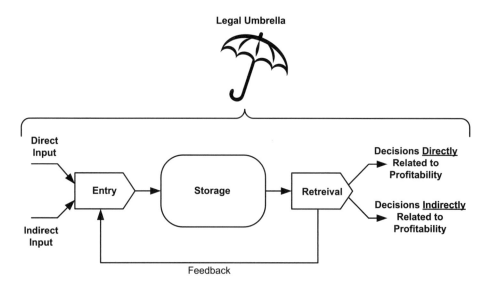

Figure 6-5 The Legal Umrella

proactively addressed before they become serious problems, companies have created internal auditing groups and have processes in place to uncover flaws. These groups monitor at various levels the working of all of the Business Internal groups to assure that they comply with the various work and reporting requirements associated with the plant's assets.

6.5   The Legal Umbrella

Every organization has retained or employed those who provide them with legal representation. Small firms usually acquire this protection through representation by a third party on retainer. Large companies typically retain legal departments of their own. These lawyers serve many purposes. In the world of asset data integrity, they have two specific functions because the vast majority of business decisions quite often have direct or indirect consequences to the business.

❏ Advising firms of potential legal issues or consequences related to business decisions (Figure 6-5).

❏ Defending actions brought against the company based on perceived negative outcomes and impacts related to business decisions.

Advising

The legal department has input to many business decisions. Some of these are through direct involvement in a specific project or work initiative where they are consulted to assure compliance with the numerous federal, state, and local regulations. However, there are business decisions made every day in which they are not consulted. This indirect involvement is handled through procedures and guidelines developed to help the company make sound decisions that may have a legal impact down the road.

It is important to note the difference between procedures and guidelines related to plant assets. A procedure is typically a document with a defined set of work steps that need to be rigidly followed in the completion of a task. Failure to execute a procedure properly can lead to serious consequences and, ultimately, legal action. On the other hand, a guideline is more of a framework within which qualified persons can perform their duties to accomplish their work. A guideline provides some level of flexibility whereas a procedure does not. This is important in the data integrity arena because both types of documents are attached to asset work packages.

A good example of this indirect involvement is the Management of Change (MOC) procedure developed so that decisions made affecting the manufacturing process are reviewed in great detail before implementation. This procedure has numerous checks and balances to assure that a decision affecting the process has been thoroughly reviewed and that everyone involved understands the impacts of the change. Although this procedure required by OSHA emerged due to serious safety issues and resultant problems, it is a clear example of a procedure that must rigidly be followed to avoid the consequences of not following it.

Guidelines are somewhat different. Take for instance a guideline that explains how to repair a pump. Although it is true that some of the steps must be performed in order, a first-class mechanic's approach to the work may differ from one person to another. This is the flexibility provided by a guideline and not by a procedure.

It is very important how these terms are used in your asset-related documentation. If it is incorrectly applied, the terminology can come back to haunt you if you become involved in an asset-related legal issue.

Another area where legal advice (direct and indirect) is provided is in the process of data integrity as data is fed into the entry portion of the data integrity transform. In this case, legal provides oversight to assure consistency and accuracy—one day what is loaded into the transform may very well be needed to defend the company against legal action.

Defending

For most companies, lawsuits are not uncommon. The outcomes of decisions that are made often have negative impacts, perceived or otherwise. In a lawsuit, a legal dispute which is most often an adversarial process, the goal of the plaintiff is to present evidence that would establish negligence or willful misconduct on the part of the defendant. The defendant is well served sometimes to rebut the evidence using objective data that would establish that the intentions were good, that requirements were met, and that the defendant acted in a reasonable manner—notwithstanding the fact that a negative impact may have occurred. Sometimes the data can help the defendant undermine the evidence of the plaintiff, thereby undermining their case.

Many lawsuits in manufacturing and the industrial world deal with negative impacts (perceived or real) to people, the community, and the environment. Defending the company often requires extensive amounts of data-related evidence. The goal is to undermine the allegations of the plaintiff and to demonstrate that the defendant (the company) has done everything possible that would have been reasonably expected to avoid the real or perceived negative consequence that has occurred. Valid data aids the endeavor.

Asset-related data about what was done, why it was done, how it was done, when it was done, and who did it are of critical importance to the defense. But what if the plaintiff can prove that your data is of poor quality, inaccurate, or—even worse—simply doesn't exist in a manageable form? In these cases, the ability to clearly demonstrate the authenticity and reliability of the data that may undermine a plaintiff's case is severely compromised. In fact, recent rulings at the federal court level have made it clear that asset-related data records may not even be allowed into evidence unless their integrity can be proven. Consider the potential cost to your company if the data you need to present in your defense is not admissible due to poor quality or simply a sloppy process for both inputting and retrieving it.

Imagine that a pressure relief valve protecting a critical process in your plant is sent out for repair. The data sheet sent along with the valve clearly specifies the set pressure (the point at which the valve will safely relieve the process if there is an operating problem). This data sheet was extracted from your asset database. However, the sheet has not been updated when the process was re-rated to a lower operating pressure. As a result, the relief valve was set too high and failed to relieve the system when it was required. The result: an explosion and serious personal injury. How well do you think the company could defend themselves in a court of law? Would the asset-related information even be allowed into evidence except maybe in support of the plaintiff's side of the claim?

The importance of the role of the legal department in the asset data integrity process can not be minimized. It is essential to obtain direct guidance where necessary, but at minimum assure that procedures and guidelines are properly in place and followed.

6.6   Indirect Aspects of the Transform

Just as the direct inputs and outputs to the data integrity transform are important to the operation and profitability of the business, so are those that have an indirect impact. Those that provide an internal impact have a great deal of influence on how departmental business decisions are made and the business outcomes of those decisions.

Chapter 7

External Outcomes and Impacts

As you consider internal impacts don't forget the ones outside of your business.
They can be significant if not addressed.

Focus Group on Asset Data Integrity

7.1   External Issues

Consider the definition of Asset Data Integrity introduced in Chapter 3:

Asset Data Integrity: A collection of points or facts about a plant asset that can be combined to provide relevant information to those who require it in a form that is entire, complete, and trustworthy.

Chapter 6 discussed the outcomes and impacts that data integrity issues can have within the organization. However, there are those outside the core business that also require asset-related data to be entire, complete, and trustworthy. No organization leads a solitary existence; when suppliers and customers are considered, one can easily see that the organization exists as a part of a larger system. There are also oversight bodies from the financial, safety and health, and environmental communities to consider, as well as the neighboring public. Asset data integrity affects decisions made by all of these groups and requires that you consider them all within the framework of data integrity. .

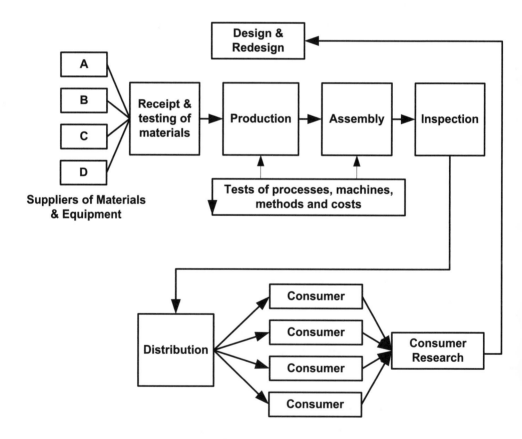

Figure 7-1 Production Viewed as a System.
(adapted from Dr. W. Edwards Deming, Out of the Crisis, page 4)

An operating plant can be considered as one part of an integrated system. Dr. Edward Deming developed this idea in 1950 during the quality revolution in Japan. He used the diagram in Figure 7-1 to illustrate this idea, and spoke of the impacts that all parts of the system have on product quality. Expanding further, maintaining operations also has an impact throughout the entire system. Quality asset data is an integral part of this system; it can facilitate increased efficiencies, enabling cost reductions throughout the entire supply chain. Like a rising tide lifts all boats, those cost reductions will occur in all parts of the system.

Whenever physical assets are operating, there is some level of risk. Whether in a hazardous chemical service or a relatively benign process, risk is present to some degree. That risk is not just internal to the organization. Suppliers depend on the continuity of orders to remain profitable. Customers depend on a reliable source of supply. Investors depend on profitability of the enterprise for investment growth and income. The public—neighbors and local communities— depend on continuity of operation for employment and maintenance of the tax base. Neighbors also depend on safe and environmentally-friendly operation. Good asset data is important to safe and reliable operation because it enables sound decision-making and affects all of the external entities mentioned.

7.2   Outcomes and Impacts—Partners

Partners are external commercial entities that have formed some sort of strategic alliance with your company. This alliance provides a synergy where the benefits for each partner will be greater than if the organizations acted alone. There are essentially two types of partners:

those that are part of the business process and those that support the business financially.

Business partners might be two companies each producing part of a finished product. Consider a plant that produces lubrication oil as a crude oil by-product. A business partner may construct a lube canning plant on the producer's site and exclusively produce the containers in which the lubrication oil is stored. This is different from a supplier because these two companies are directly linked; the producer of the cans is providing them solely to the lube manufacturer.

Financial partners provide operating cash to a business with the expectation that they will enjoy part of the profit. This often happens when a person wants to create a company but needs an infusion of cash. Rather than execute a bank loan, they take on partners (silent or otherwise) to help fund the endeavor.

Partners derive benefits from accurate asset data in a number of ways:

❏ Accurate data yields sound information and good value-based business decisions that foster partner trust. This level of trust is critical to any partnership; the belief that each partner will uphold their end of the bargain builds longevity in the relationship. The opposite is also true. The loss of trust is the quickest way to ruin a partnership. Although there are numerous ways that this loss can occur, poor asset data leading to faulty decisions is certainly one way.

❏ In the event of a merger or acquisition, good quality asset data enables accurate valuation of the physical asset base and facilitates the "due diligence" process, resulting in a fair price paid for the assets. This aspect is important for the partnership because due diligence and accurate asset valuation enable the partners to obtain fair value. This aspect also applies in the case where one partner may want to buy out the other in the partnership.

7.3   Outcomes and Impacts—Suppliers

Supplier relationships with the plant can be dramatically enhanced through reliable asset data. MRO (Maintenance, Repair, and Operating) equipment and material suppliers are more likely to provide good service when the exact equipment or specification of equipment parts are known and accurately provided. Consider the example of a set of heat exchanger gaskets. These gaskets were ordered by the maintenance planner for a job that was not to take place for two weeks. This gave the supplier sufficient time to construct them based on the specification drawing provided. The gaskets were delivered in plenty of time, but when they were installed they did not fit. The reason: the company had changed the gasket surface dimensions, but never updated the asset-related data.

The effect on the supplier was that they had to work through the weekend around-the-clock to provide a new set of gaskets for the customer. Although the supplier's workers made extra money, it just happened to be a holiday that they would have preferred spending with their families. After this type of problem occurred several times, the supplier lost trust in their customer and went to extreme lengths to create their own gasket files. Certainly this isn't the way to address this issue. Accurate asset-related data would solve this problem for both parties versus the negative impact that they both experienced.

Suppliers also can benefit from improved asset-related data by taking advantage of the current technology. Electronic Data Interchange (EDI) arrangements can be implemented that eliminate the need for phone calls and paper purchase orders, reducing transactional costs. The vendor base can be consolidated, allowing fewer vendors to share in a larger slice of the business. Depending on the geographic proximity, there is also the opportunity to bundle purchases from several sites within the same company into a single purchase order. Bundling results in fewer orders, a larger dollar value per order, and fewer invoices to manage, reducing handling costs.

Other types of suppliers can also benefit:

❏ Architect / Engineering firms benefit from accurate data by having the capability of knowing the specific assets in the plant. When providing designs for plant expansions or new system installations, they have access to capacity specification detail about the equipment. They can assist the plant with equipment standardization efforts by specifying similar equipment to the installed base. The cost to set initial maintenance strategies and to create bills of materials is reduced by the ability to leverage the work done on similar or identical assets in the plant.

❏ Contractors benefit from accurate data by gaining an increased ability to plan their work. Improved planning will result in labor efficiency gains that will improve profitability in fixed price contracts. For outage or shutdown work where the contractor is providing supplemental labor for repair or overhaul of equipment, engineering drawings and manuals can be obtained in advance because the specific equipment make and model are known. The ability to pre-fabricate subassemblies is enhanced because details such as dimensions, clearances, flange sizes, and other fit details are identified.

❏ Repair shops indirectly benefit from accurate asset data. Equipment specification detail may open opportunities for local machine shops to fabricate spare parts, reducing dependence on OEM (original equipment manufacturer) parts. Rotating spare equipment such as electric motors and pumps are more easily tracked with good asset data. This also applies to relief valves, which are often spared and rotated into and out of service. The repair facility can increase service by keeping data concerning repair scope and frequency; it can use this data to inform the plant of "bad actor" equipment. This notification is particularly important for electric motors that can experience efficiency degradation caused by frequent rewinding.

7.4   Outcomes and Impacts—Customers

The customer value proposition is the total benefit that a customer receives in return for the associated payment. Maintaining reliable and accurate asset data increases this value proposition:

❏ As seen in Chapter 6, accurate data helps provide the foundation for greater manufacturing reliability, increased efficiency of production, and lower manufacturing costs. From the customer's perspective, this translates into lower prices and increased reliability of supply. After all, the supplier / customer relationship is almost the same as the partnership example. However, in this case, the customer can obtain goods from a vast array of suppliers. If you were in the customer's position, wouldn't you rather obtain your goods from a supplier that ran an effective and efficient process? Although asset data integrity is not the only component, it certainly is one of the major ones.

❏ Some customers require purchased goods to be traceable. According to the ASME Boiler and Pressure Vessel Code, Section III, Article NCA-9000, "Traceability is the ability to verify the history, location, or application of an item by means of documented recorded identification". Good quality asset data enhances traceability by providing additional detail concerning the manufacturing history of the product.

❏ High quality asset data can ultimately result in greater customer confidence. If suppliers have the discipline to maintain good, accurate data, they are more likely to have the discipline to do the other things well that result in quality products all the time.

7.5   Outcomes and Impacts—Agencies

Regulatory agencies also rely on asset data to verify that a facility complies with applicable regulations. There are common requirements for data regardless of the regulatory body, industry, or regulatory law. Considering the emphasis that these diverse agencies place on asset-related data integrity, it is very clear they are firmly convinced that the maintenance of this data is critically important to the safe, reliable, and environmentally sound operation of industry. Some examples of regulatory agencies and their requirements for asset data are as follows:

Occupational Safety and Health Administration

The Occupational Safety and Health Administration was created in 1970 with the responsibility for creating and enforcing health and safety regulations. Several OSHA regulations require employers to maintain specific information concerning physical equipment assets. One of the more stringent regulations is the Process Safety Management (PSM) Law, OSHA 1910.119.

PSM was formulated to govern companies who produce, use, or store highly hazardous chemicals. Its major objective is to put a process in place that will prevent unwanted releases of hazardous chemicals—especially into locations that could expose employees and others to serious hazards. Process safety management is the proactive identification, evaluation, and mitigation or prevention of chemical releases that could occur because of failures of processes, procedures, or equipment. Specific requirements of this law pertaining to physical asset data include:

❑ The design basis for the equipment must be documented to ensure that the equipment conforms to Good Engineering Practice and is fit for the service in which it is installed.

❑ Specific information (referred to as Process Safety Information) about the equipment such as materials of construction, electrical classification, capacities, etc., must be maintained and up to date.

❑ The mechanical integrity of the equipment must be maintained to ensure that equipment failure does not result in a release of hazardous chemicals. This requirement includes ensuring that spare parts used in maintenance or repair—such as bolts, gaskets, seals, and other parts—are suitable for the service in which the equipment operates.

❑ Any changes to the equipment must go through a Management of Change analysis to ensure the changes will not result in an inadvertent chemical release. When complete, the Process Safety Information must be updated to reflect the change.

Environmental Protection Agency

The Environmental Protection Agency (EPA) was formed in 1970 for the purpose of protecting human health and the environment. It has the responsibility for setting and enforcing national standards under a variety of environmental laws. The EPA also has several regulations requiring companies to maintain specific information concerning equipment.

❑ The owners or operators of equipment in a service in which a release could harm the environment are required to maintain Process

Safety Information about the equipment. Specific information includes materials of construction, design basis, the design standards and codes involved, and any safety systems or interlocks that will prevent a release.

❏ A Quality Assurance process must be in place to ensure that the equipment as it is fabricated is suitable for the process application for which it is used.

Nuclear Regulatory Commission

In 1991, the Nuclear Regulatory Commission promulgated a regulation that requires operators of nuclear plants to monitor the effectiveness of maintenance activities on systems and equipment that are critical for safe reactor operation. Known as the Maintenance Rule, this regulation seeks to ensure that "systems, structures and components" whose failure could cause an accident resulting in offsite exposure, are required for safe shutdown, or are required to mitigate accidents remain capable of performing their intended functions. In response to this rule, the Nuclear Utility Management and Resources Council (NUMARC) issued a guidance document intended to assist nuclear power plant operators in implementing the Maintenance Rule. This document contains provisions related to maintaining asset data:

❏ The systems, structures, and components (equipment) covered under the provisions of the Maintenance Rule and the criteria for inclusion shall be documented.

❏ Activities associated with the preventive maintenance program—the results of repairs, tests, inspections, or other maintenance activities should be documented.

❏ Documentation of systems, structures, and components subject to American Society of Mechanical Engineers (ASME) code testing should be maintained.

Food and Drug Administration

The U.S. Food and Drug Administration (FDA) is responsible for regulating and supervising the safety of foods, tobacco products, dietary supplements, prescription and non-prescription medication, vaccines, biopharmaceuticals, blood transfusions, medical devices, veterinary products, and cosmetics. The FDA has issued guidelines to ensure that products are manufactured safely and that the products maintain high standards of quality. Referred to as Good Manufacturing Practice (GMP), these guidelines govern all aspects of production that would affect the quality of the product. Although separate guidelines are issued for the different industry segments listed above, common requirements regarding equipment design, care, and maintenance are facilitated by sound asset data:

❏ A written record of major equipment cleaning, maintenance, and use shall be included in individual equipment logs.

❏ All plant equipment in food contact service shall be properly maintained and maintenance records must be kept.

❏ Documentation of calibrations must be maintained indicating the instrument that was calibrated, the date of calibration, the reference standard used, the calibration methods used, and the calibration readings found.

Sarbanes-Oxley Act

The Sarbanes-Oxley Act is a law that set new or enhanced standards for financial disclosure for all U.S. public company boards. Enacted in 2002, it was developed as a reaction to a number of major corporate and accounting scandals that involved securities fraud. It led the U.S. Securities and Exchange Commission (SEC) to adopt several new rules concerning auditor independence, corporate governance, assessment of internal controls, and enhanced financial disclosure.

Although the Sarbanes-Oxley Act does not contain specific language concerning asset data, it requires the principal executive officers of public companies to certify that they have presented the financial conditions and results of operations of the company fairly and accurately. Because property and equipment typically make up a significant portion of the value of a company's asset base, accurate asset data enhances the accuracy of the company's balance sheet. Property accounting records can be reconciled against the master equipment list to ensure that the reported asset values are correct.

7.6   Outcomes and Impacts—Public

A great many manufacturing plants are located in close proximity to residential communities. Quite often if there is a plant problem resulting in a major fire or release, there is community impact. Thus, the public has a vested interest in asset-related data quality because the lack of this aspect of the business process may have an adverse impact on their very lives. The public gains an indirect benefit from good quality asset data in a number of ways:

❑ The investor public benefits from greater accuracy in financial reports as noted above. Individual investors are more capable of assessing the value of the business and deciding whether it would be a good investment.

❏ The community in which the business resides benefits because good asset data provides an accurate representation of plant and equipment values for property tax purposes.

❏ The Emergency Planning and Community Right to Know Act was enacted in 1986 to help local communities protect public health, safety, and the environment from chemical hazards. Provisions in this act help increase the public's knowledge and access to information on chemicals at individual facilities. Under this act, state and local governments are required to create emergency response plans to address any incidences of chemical releases. One of the requirements of the emergency response plan is that a training program has to be developed to provide emergency responders with information about the equipment and chemicals that they will encounter in the event of a release. An accurate and detailed equipment inventory will assist in the effectiveness of this training plan.

7.7   Outcomes and Impacts—Insurance Carriers

Insurance is a form of risk management used to hedge against a risk of loss. It can take many forms, but for the purposes of this discussion, we will consider a typical industrial business owner's insurance. Most organizations carry insurance policies against the risk of significant property damage, business interruption caused by fire, equipment failure or other catastrophe, and liability protection. In essence, the risk of the potential event is transferred to the insurance company in exchange for a premium. The cost of these policies varies according to the probability of the loss and the financial value of the policy.

Insurance companies may inadvertently find that the insured organization behaves somewhat differently than if it were fully exposed to the risk—by definition, the client organization has transferred the risk to the insurance company. For example, an organization may decide to defer maintenance tasks on critical equipment to save money, which magnifies the risk of an equipment failure. In such cases, the insurance companies may have a clause in the contract that gives them the right to inspect compliance to the maintenance program and to charge higher premiums if they discover that the client has not performed the tasks as scheduled. In this situation, the insurance company directly benefits from good asset data and the historical record of maintenance performed because it can quantify the probability of loss.

7.8   The External Impacts Are Important

Quite often when we work within a processing plant environment, we believe that the plant is the system. As you have seen in this chapter, this is not the case. The plant is a part of an integrated system composed of suppliers and customers at the opposite ends of the value chain as well as many other intermediaries that are directly or indirectly involved. Although asset-related data integrity is not the only aspect that links all of these parts together into a manufacturing system, it is a significant part. If this phase of the business is handled correctly, it almost goes unnoticed by those within the system. However, if handled poorly, the negative aspects to all of those involved can be severe.

Chapter 8

Information Technology (IT) Problems and Solutions

> IT resources are supporting functions in the world of data integrity, and like all things a structure can't stand without good support.
>
> *Data Integrity Work Team*

8.1   The Implication for IT

Throughout this book, we have discussed various aspects of data integrity that touch or are touched by the world of Information Technology. The extent of the involvement of IT when it comes to issues of data integrity is large and far reaching. Things like computers and network infrastructure, software applications, database architecture, and connectivity challenges, all relate to data integrity and are managed by IT in most organizations. In this chapter, we will try to paint a picture of the various models of modern asset management practices, and how they both impact and are impacted by the information technology resources of a company.

8.2   Implications to IT of a Modern Asset Management Practice

In Chapter 9, we discuss an enterprise-level data integrity model and suggest that asset-related data will reside in a variety of applications. However there should be a single repository of all asset-related data that is used by all applications. Furthermore, we believe that the CMMS or EAM system (or the maintenance and asset management modules of the Enterprise Resource Planning (ERP) system) should be the repository for asset data. All other systems requiring asset data should take their asset master data from this single master source. We caution against keeping duplicate asset data in multiple systems. More about this in chapter nine.

Suppose a specialized vibration analysis software application requires asset information as master data. The CMMS system, where we suggest the asset master data should reside, does not and likely will not ever be able to perform the very specialized and sophisticated analysis that only vibration analysis software can perform. But if we want to keep our master asset data in only one place—and specifically in the CMMS system—then the implication is that the asset master data needed by the vibration software should be charged or fed from the CMMS system.

8.3   The Advent of ERP Systems

As recently as 10 years ago, during the advent of Enterprise Resources Planning systems like SAP, the desire began to move from a myriad of specialized and unconnected software applications to an integrated, multi-purpose software application. The idea was twofold. First, the presumed advantages of all applications operating in a single system using a single consistent database would solve the problem of multiple and sometimes disagreeing sources of data (asset data in our case). The inefficiencies of having multiple sources of the same information in various applications, and not knowing which source was the truth, were to be solved.

Second, the simplification of the information technology function, including simplifying and minimizing the number of software applications and databases needed, was expected to significantly reduce the costs of maintaining and operating multiple software applications and databases. The need for various skill sets would presumably be simplified.

The movement toward ERP systems by most major corporations ensued and a lot of these advantages were realized. Today, most Fortune 500 companies have replaced their major business applications with a single ERP system—with all business applications (e.g., finance and accounting, supply chain management, human resources, logistics, manufacturing control, customer service, billing) integrated into a single suite of software.

One business function that has lagged others in terms of being included in the ERP model is the maintenance or asset management function. During the early days of ERP systems, the functionality of the ERP's maintenance modules was very weak compared to the specialty CMMS or EAM systems they were designed to replace. As a result, many companies opted to acquire or keep a stand-alone maintenance system and interface it, using custom interfaces, with their ERP system. Over time, the functionality of the ERP maintenance modules has increased; they are now comparable to the stand-alone systems. Many companies are now deploying the maintenance modules of their ERP system and discarding their stand-alone CMMS or EAM systems.

More recently, the largest ERP software companies, including SAP, have recognized that they have incorporated most major business functions into their integrated application suites, and yet there are many specialty software applications that continue to operate and coexist. The ERP companies have steadily added more and more business functions in an attempt to be the single application suite used in a client company. Asset management and maintenance has been a particularly fertile area, often with a landscape of dozens or more specialty software applications in use at typical industrial companies.

But will every specialized software application be replaced by the more complete ERP of tomorrow? Is it possible for an ERP software company to have the expertise and resources to be able to do everything related to assets? Even if it would be possible, is it advisable and practical? Can the innovation and creation of narrowly-focused solutions for asset management be accomplished by one big ERP company or will smaller specialty entrepreneurial companies continue to invent solutions for new asset management problems?

We cannot predict the future. Although it is true that the breadth and scope of functions solved by the ERP systems has increased steadily over the past 10 years, we suspect that not all asset management applications will ultimately reside within the ERP system. Observing the marketplace, it is apparent that the ERP companies are seeking to collaborate with certain specialty players in the marketplace. Some of these specialty players are quite large and established. For example, most industrial companies already have deployed distributed control systems (DCS) in their industrial plants. There are well-established global companies that provide these solutions and will likely continue to do so. It is unlikely that the ERP software companies will want to get into the DCS business.

But as asset management practices continue to advance, we now know that the information contained in these DCS systems—which largely deal with physical parameters of the industrial processes like temperatures, pressures, and flows—can be very useful in better understanding the current capability of a process and the assets that make up the process. Some companies are inventing software applications that take advantage of the data in the DCS systems. They are attempting to combine this data with the asset data in the CMMS system to predict the likelihood that a set of assets will be able to produce a product on a specific schedule. The niche of predictive analytics takes advantage of all data from many systems to increase the ability to understand and predict asset and process capabilities.

Many other examples exist of data sources that could be combined with other data and used to improve the predictability of asset performance. These will most likely continue to coexist in a typical industrial company. From an IT perspective, one of the emerging challenges will be to figure out how to reconcile the desire to simplify and minimize information technologies deployed within an enterprise with the reality that many specialty software applications will continue to exist outside the ERP system.

IT resources will help their organizations decide which business functions should reside inside the ERP system, and which ones should or could not do so. They will also have to facilitate the simultaneous use of multiple asset management systems and the management of the master data held inside each of these systems. They will have to develop connections and interfaces that allow holding of asset master data in only one place and feeding that data to other applications needing such master data.

There are obvious technical challenges to be overcome by IT

inherent in the modern asset management practice. But at the heart of this challenge is data management. Devising and then managing a data model that can be used by all applications—and used to properly marry the data from one system with the related data from other systems—is a goal that must be solved before the ideal model of asset management can be fully realized.

For example, the DCS data is organized and stored by tag number. But these tag numbers have no relation to the asset numbers in the CMMS system. If a data model existed which made it possible to marry the tags in the DCS databases with the asset structure in the CMMS system, a fuller set of data can be viewed in one place and analyzed to better understand the current and future capability of a set of assets to produce a product. Likewise, the example we used earlier of the specialty vibration analysis software reminds us that narrowly-focused, highly-specialized software applications will likely never be replaced by the ERP system. Therefore, figuring out a data model that allows the asset information shared by these multiple systems to be housed and maintained in one single system, and then served up to the constituent consumers of that information, is an IT challenge, and one worth solving.

8.4   Master Data Management

For several years, a cottage industry called Master Data Management has evolved; this industry deals with the formatting and maintaining of master data. Recall from earlier chapters that master data is reference information about subjects of interest. In our case, the identification and physical descriptions of the assets is master data, distinct from transactional asset data like current asset condition, future asset capability, etc. This master data fundamentally describes the assets and is called upon by various aspects of the software application. The master data is maintained in one place and is served up to be used by the various functions of the software applications.

Master Data Management has primarily been focused on customer data—that is, master information about a company's customers such as their names, addresses, and contact information. This master data is maintained in Customer Relationship Management (CRM) software applications. Niche companies and specialty master data management software tools have appeared, all providing the vital service of ensuring that the master customer data is correct, accurate, complete, consistent, etc.

Again, from an IT perspective, we believe that the migration of this cottage industry into the realm of assets has begun; it will develop and continue to mature. IT resources will likely be at the center of ensuring that Master Data Management for assets and asset-related data is centrally controlled, consistently formatted, and maintained to be current as changes in the actual plant trigger the need for master data changes.

8.5   The Future

This chapter has provided some insight into the future landscape of an ideal enterprise asset management model and how that landscape will challenge the IT resources of our companies. This is not an instructive chapter for IT professionals about the technical aspects of information technology as it relates to asset data integrity. But we hope the description of what is likely necessary for asset managers at an enterprise provides the insight into the IT challenges of interconnectivity of applications and the data that resides in those applications. Needless to say, data integrity in any company will depend significantly on the information technology resources supporting, facilitating, and in some cases managing the enabling components that will be necessary constituent parts of a robust data integrity model.

PART TWO

Building a Sound Data Integrity Process

At this point the problems associated with asset data integrity are more than recognized. But recognition is only a first step of the journey. The question that must be asked and answered is, "Now that I recognize the problem, what do I do about it?" The answer is contained in Part 2.

Chapter 9

Building an Enterprise-Level Asset Data Integrity Model

> Enterprise thinking proves that the sum is more powerful than the individual parts.
>
> *Data Integrity Work Team*

9.1 Historical View

Traditionally, physical asset data has best been managed at a plant or facility level, vs. an enterprise level. Most companies relegate the management of asset data to various responsible parties within a plant. They fail to recognize the strategic importance of asset data integrity in achieving corporate business goals, whatever they may be. Within a typical industrial plant, there is not one single asset data custodian. Instead, there is more than one, each responsible for different classes of asset data.

For example, data for electrical distribution equipment, or instrumentation and controls, do not typically reside in the plant's Computerized Maintenance Management System (CMMS) or

Enterprise Asset Management (EAM) system, where rotating machines like pumps, fans, and compressors may reside. Stationary equipment like tanks, vessels, valves, piping, heat exchangers, and reactors may reside in yet another system altogether.

From the isolated perspective of any one of these individual data custodians, it is easy to justify separate data management based on unique and different activities, attributes, and uses of their respective data. For example, instrument technicians often express the need to manage their asset data separately because of special calibration activities that are not typically supported by the leading CMMS or EAM systems. These interests are well-placed and understandable in light of the isolated view of each group, and in light of the often-misunderstood perception that their interests cannot be met in a more centrally-controlled and managed asset data integrity model.

In fact, with a properly managed and standardized data integrity model, the interests of the overall company, as well as the interests of the special interest groups, can all be met. Once the strategic importance of asset data integrity is fully appreciated, significant benefits at the plant level are possible. The biggest benefits come from the elevation of asset data integrity management to the corporate or enterprise level. We will explore this concept in this chapter.

More specifically, in this chapter, we will discuss in some detail what it takes to build and sustain a sound asset data integrity model, both at the plant level and at the enterprise level. The following topics will be covered:

❏ What is an asset, the elements of asset-related data, and asset classification

❏ The difference between static data and dynamic data

❏ The differences among an asset, a functional location, and func-

tional location hierarchies
- ❏ Other asset-related master data
- ❏ How asset-related data elements and attributes should be structured, organized, described, and formatted
- ❏ Ideal asset data repositories
- ❏ Enterprise-level vs. plant-level asset data integrity

9.2   What Is an Asset?

When we are talking about asset-related data, just what are we talking about? This fairly complicated question is difficult to answer simply. Generally, the assets we are referring to are those physical equipment items that make up the manufacturing, industrial, and facilities systems, processes, units, and lines that fulfill the mission for which they are intended. For example, a large building has a lot of equipment (assets) designed to provide heating, ventilation, air-conditioning, lighting, plumbing, and other functions that make the building habitable and comfortable. A more involved example is a power plant that has a lot of equipment designed to produce and distribute electricity.

For purposes of this book, we are using the words *asset* or *physical asset* to refer to the constituent components of the general term *equipment*. But what do we mean by *asset*? We start by choosing an example that most readers can relate to regardless of industrial background: your automobile. The engine of your car can certainly be described as equipment—or just as easily as a physical asset. So is the entire engine an asset then? Not really. The engine in a typical automobile is a fairly complicated system made up of many sub-systems and components. For example, an internal combustion engine (the most prevalent type of automobile engine today) has an ignition system, a fuel system, a cooling system, a combustion system, an exhaust system, etc. Is each of these systems an asset?

In turn, each of these systems can be further broken down into constituent components. The ignition system might have spark plugs. The fuel system might have fuel injectors and a fuel pump. The cooling system might have a water pump. The exhaust system might have a catalytic converter and a muffler. You get the idea. Are each of these components assets?

Again, couldn't these also be broken down further? For example, the water pump may have an impeller, bearings, a shaft, a pulley, etc. Are these assets? The pulley is made of a metal that starts out as iron ore and is melted then molded or cast into the pulley. Is the metal an asset? The pulley is affixed to the shaft using bolts. Are the bolts assets?

The answer generically is yes to all of these. But in the discipline of physical asset management, the "nuts and bolts" are not maintained. They are instead discarded and replaced when they fail. Therefore, they are not considered assets. They are considered spare parts. The deciding factor when determining whether a component should be considered an asset should be based on whether the component is maintained and repaired (vs. discarded and replaced when failure occurs). If we can repair it and restore it to service, it is generally what we call a maintainable asset. If, on the other hand, a component would be discarded and replaced, it would be referred to as a spare part (and not an asset). Another way to look at this categorization is that if the component is at the level of granularity where the next level down could be considered a spare parts list, then the component is a maintainable asset we refer to as a physical asset.

Although this approach governs most decisions about what to call an asset, there are some exceptions. Other factors which might designate a component a maintainable asset include:

❏ A component for which we want to track history of failures, repairs, parts usage, etc.

❏ A component which requires special record keeping—for example, regulations require cranes and other hoisting devices to be tested periodically; therefore, cranes and hoists would be considered maintainable assets, regardless of whether they are repaired vs. replaced.

❏ A component which may rove from place to place during its life, and the tracking of where it is today (as well as where it has been over time) is important.

❏ A component which, no matter how small or repairable (vs. replaceable), is highly critical to the mission. An example of this is the car battery in an ambulance. Although this is replaced when it fails (vs. repaired), it is critical to the mission of the larger system. Thus, we would want to track its history, monitor its charge and usage, etc., to ensure that we don't find ourselves stuck with no ability to transport a critically-ill patient to the hospital.

9.3 Asset Classification

Once we identify our Assets, they can be categorized into general asset categories describing the type of asset. The following sample asset categories list examples of assets that fall into that category:

• **Rotating Assets** (pumps, compressors, fans, blowers, motors, turbines, generators)

• **Stationary or Fixed Assets** (tanks, pressure vessels, heat exchangers, valves, piping)

- **Electrical Assets** (transformers, switchgear, breakers, motor control centers, starters)
- **Instrumentation and Controls** (indicators, gauges, sensors, transmitters, controllers, alarms, valve actuators)

Assets can also be categorized separately by the kind of functions they perform, such as Fire and Safety, Production, Utilities, Security, Environmental Protection, and Packaging.

At the asset level itself, each asset can be classified by asset Class and Sub-class. The following are examples of asset Classes and Sub-classes:

❏ Asset Class = Pump; Asset Sub-classes = Centrifugal, Reciprocating Diaphragm
❏ Asset Class = Motor; Asset Sub-classes = AC, DC

Asset class and sub-class are important in that they are used by most asset information software systems to define various attributes of the database record that describes an asset. For example, a centrifugal pump has physical characteristics that describe it which are different than those of a motor. Attributes of a centrifugal pump might include size, shaft diameter, material of construction, impeller diameter, inlet connection size, outlet connection size, capacity in gallons per minute or feet of head, manufacturer, model number, and serial number, to name a few. Attributes of an electric motor might include horse power voltage, amperage, frame size, rotations per minute (RPM), service factor, insulation class, manufacturer, model number, and serial number, to name a few.

Notice that there are some common attributes between these two asset classes (manufacturer, model number and serial number), but the other attributes differ from class to class. So classifying each asset allows the development of a format for attributes that is unique to that asset class and sub-class. Most asset information systems allow for the creation of a record in a database that describes the asset; they provide a field in that record to allow for the entry of the Class and Sub-class. That assignment then triggers which fields of data are needed for that specific class of asset.

Asset classifications also trigger which failure modes would appear on a pull-down list as a worker is closing out a repair work order on that asset. The failure mode list is "context-sensitive" to the classification, presenting only those failure modes that relate to that class of asset. Thus, asset classification is very important in modern asset management strategies.

It might be noteworthy here to refer to some emerging standards that provide guidance on the development of an asset structure in a CMMS or EAM system. The International Organization for Standards (ISO) is a worldwide federation of national standards bodies, and various technical committees develop standards for various purposes. Technical Committee ISO/TC 67 deals with Materials, equipment, and offshore structures for the petroleum, petrochemical, and natural gas industries. This technical committee originally created ISO 14224 back in 1999—that standard has been updated in 2006 based on lessons learned from the use of the previous version. ISO 14224:2006(E) attempts to provide a comprehensive basis for the collection of reliability and maintenance data in a standard format for equipment in the facilities and operations of the industries of interest. This standard is being used by many companies, both in and outside of the petroleum and petrochemical industries as a guideline for the formatting and structuring of asset-related data.

Figure 9-1 shows a typical asset record screen for a pump, including information that is entered when the asset record is created. Figure 9-2 shows a similar example for an electric motor.

9.4   Static Data vs. Dynamic Data

As you can see in Figures 9-1 and 9-2, the information, or data, that is displayed is mostly descriptive information about the asset. Information which describes the physical characteristics about the asset that tends not to change over time is considered static data. Conversely, information about an asset that has the likelihood of changing frequently as time goes by is referred to as dynamic data. In most cases, static data is created once and is then appropriate (accurate and complete) for long periods of time. Static data is also known

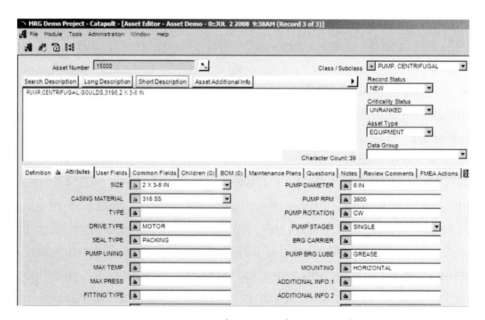

Figure 9-1 Asset Record Screen for Centrifugal Pump

as Master Data.

Information about an asset that has the tendency to change over time—dynamic data—may be entered (populated) automatically by the software application by virtue of transactions occurring, e.g., work orders. Or dynamic information about an asset has to be manually updated when physical changes occur, in order to keep the data about the asset in synch with the actual physical configuration of the asset.

For example, typically the processing of a work order for repair of an asset would automatically record in the equipment history area of the asset record the fact that the work was in fact performed, when it was performed, who performed it, how many hours it took to perform, what spare parts might have been replaced during performance, etc. This kind of dynamic data is generally created automatically through the course of transactions taking place in the CMMS system. So the

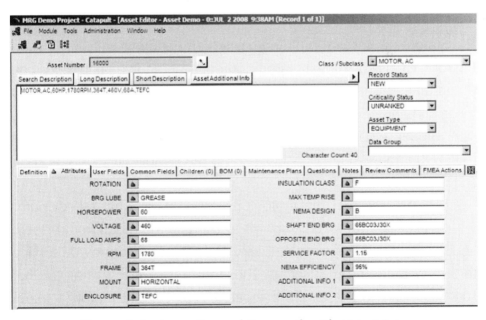

Figure 9-2 Asset Record Screen for Electric Motor

configuration of the transactional data in the asset record always stays accurate and complete, assuming of course that the work order functionality of the system is being used properly.

Some dynamic data tend to require manual intervention in order to keep it accurate and up to date. For example, the location where the asset is installed in the plant or the building may be accurate when that information is initially populated. But roving assets can move to different locations in the plant or building. Sometimes the data about that asset has to be manually updated when the physical asset moves. Many asset management information systems do not automatically update the asset record with its new location, although some do.

In the older asset management systems, the kind of information that described the functional location where the asset is installed, including the function the asset is called upon to do in that location, resided in the same asset record where the static descriptive data about the asset resided.

Suppose a hospital building has ten air handling units (AHUs) on the roof, each providing air conditioning, heat, and ventilation for one of ten different operating rooms. Assume that the assets that make up those air handler units are all the same type (same manufacturer, type, size, model, etc.). Each of those air handler units has a main drive motor as a maintainable asset— again, the motors vary only in their serial number.

Let's assume that the motor on AHU#1 fails. That failed motor would likely be removed, sent off to be repaired/rebuilt/refurbished, and then put in storage for later use. AHU#1 needs to provide service to the operating room it serves. Therefore, instead of waiting for that specific motor to be rebuilt, a separate replacement motor (a spare) would be installed in its place. The function of the air handler unit for that operating room could then continue while the original motor was being rebuilt. At some point, the failed motor is rebuilt and put into storage as a spare, ready for future use.

Continuing this example, assume at some time in the future AHU#5's motor fails and the refurbished motor that was taken from AHU#1 is installed into AHU#5.

To keep the configuration of the asset data accurate over time, a data integrity strategy would have to have been employed, with a number of options being viable. It is of interest (and best practice) to want to understand the actual repair history (cradle-to-grave) of that specific serial number motor asset. Therefore, the asset record for the motor should stay with the physical motor wherever it resides—vs. stay with the functional location. So in this scenario, our original motor from AHU#1 would had to have been modified to indicate first that the motor was now in stock and not installed in any AHU functional location, and eventually modified again to indicate that it was now installed in AHU#5 service.

A different and frequently-employed strategy, which is not considered best practice, is to have the motor asset record represent the functional location. In our example, this location would be as the motor for AHU#1. In this scenario, we don't address the fact that the physical motor asset may move around. Instead, the new motor installed in AHU#1 service assumes the identity of the previous motor asset record, including the history of the previous motor. Unfortunately, you lose the cradle-to-grave history tracking of the specific motor that is there now and the one that was there before.

The reason is they eventually assume the asset record (and the history, etc.) of another location's motor. The person responsible for asset data integrity would have at least had to change the serial number of the motor asset record when the new motor was installed in the original location, but again the cradle-to-grave history is lost. What is the best practice to deal for these realities? Read on please.

9.5   The Differences Among Assets, Functional Locations and Functional Location Hierarchies

In reality, either of these scenarios is not best practice. Furthermore, the most common practice is that the proper data integrity maintenance is not done at all. The configuration of the asset data is very often out of date and not accurate because these movements of roving assets are not addressed properly, if at all.

As a response to this reality, most of the top-tier asset management systems today provide functionality to create a parallel table in the system. The table is populated with records representing the functional locations within a plant or building. A separate table is populated with records representing the physical assets that may be installed in those functional locations.

When an asset record is created, it is assigned to a functional location record based on where it is today. In this model, the functional location records are populated with the information about the location. NO information about the physical characteristics of the physical asset currently located there should be entered into the functional location record. Conversely, the asset record would be populated with only information about the physical characteristics of the asset, with NO information about the functional location. In this way, when the asset moves to a new location the data integrity maintenance is minimal. The only thing that has to change is the functional location assignment field in the asset record.

Some of the top-tier asset management systems provide automatic functionality during the work order process that asks the user to enter the new location number for the asset if it was moved. This functionality makes the data integrity maintenance task easier, maintaining the currency and accuracy of the configuration of the asset data.

In this model, we can have our cradle-to-grave history of physical assets regardless of where they have been installed over time, as well as the historical information about the functional location, including which physical assets have been installed there over time. This model minimizes data integrity maintenance and enhances data integrity, and is considered best practice.

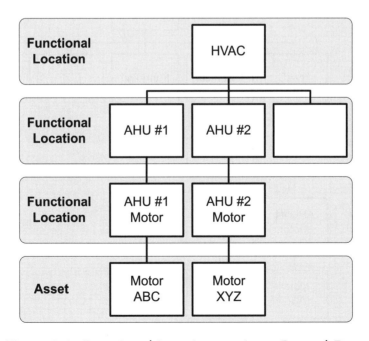

Figure 9-3 Functional Location vs. Asset Record Data

Figure 9-3 describes this concept using our hospital air handling unit example.

One final thought on the functional location vs. asset concept—the records in the functional location table can be related to each other in a hierarchy that organizes the plant in a logical way. Figure 9-4 provides an example.

As you can see from Figure 9-4, our hospital sits at the top of the hierarchy, with the next level down being the various systems in the hospital (including Heating, Ventilating and Air Conditioning (HVAC), Electrical Distribution, Fire Protection, Lighting, Plumbing, etc.).

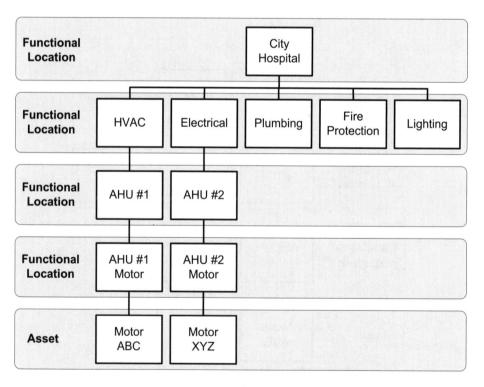

Figure 9-4 Functional Location Hierarchy

Under HVAC, our Air Handler Unit functional location records appear, and under them the Drive Motors for those AHUs. At the bottom of this functional location hierarchy, we see that the physical asset records are assigned, shown in Figure 9-4.

Notice the functional location records are described in terms of that function or location (and not in terms of the physical characteristics of the assets that may be assigned there). Likewise, the asset records are described only in terms of their physical characteristics (and not in terms of the functions or locations they may be installed in at any given time).

This model is considered best practice and eases the maintenance of the data integrity. It also allows other functionality to occur, including providing drill-down searching capability when looking for assets in the database, and providing the roll-up of costs through the hierarchy so that costs of maintenance for all assets below the chosen functional location can be collected.

9.6   Other Asset-Related Master Data

Aside from the asset records and the functional location records, there are other tables that have to be populated in most asset information systems that are considered master data. A few examples follow (generic terms are being used here—various asset management systems have specific names for these codes and tables):

❏ Asset classes and sub-classes, including attributes for each combination
❏ Cost centers
❏ Asset criticality codes
❏ Work Order priority codes

❏ Work type or work class codes
❏ Failure codes (types, modes, causes, remedies, etc.)

In most cases, these master data tables have to be populated in order for a pull-down list of choices to appear when prompted to fill in that field. Careful thought and strategy must go into devising the entries in these tables, taking into consideration best practices of asset management and the functionality of the specific asset management software system in use.

9.7   Asset Master Data Structure and Formatting

Is it possible to establish rules and strategies for the structuring and formatting of this asset-related master data? Are there best practice standards that guide such rules? Is it possible to standardize centrally across all of the interest groups in the plant, or even at the enterprise level? Let's take each of these questions one at a time.

Is it possible to establish rules and strategies? Absolutely, and it is essential if the full entitlement of benefits is to be achieved. In fact, as we discussed in Chapter 3, the master data management arena has adopted the term taxonomy to describe such rules. Taxonomy is a comprehensive data structure that permits consistent classification of any person, place, thing, or idea managed by a system.

Note the word consistent within the definition of taxonomy. The actual definition assumes that consistency is one of the drivers for a taxonomy. Is it possible to have a taxonomy for asset-related data that can be adopted by all of the various interest groups within a plant who presently manage silos of asset data? Again, the answer is an emphatic yes. Furthermore, it is not just possible, it is also necessary for alignment with best practices thinking. Let's put the word taxonomy into more specific terms related to asset management.

Think of a taxonomy as a rule-set, or a template, or a data entry form into which you enter your specific data. Remember the asset class and sub-class are typically assigned to an asset record to classify it and trigger specific fields of attribute data that we want to record about that specific class of asset. Recalling our example, a Pump is a common class and Centrifugal is a common sub-class of Pump.

Having a list of choices from which to choose the appropriate class and sub-class requires that list to be constructed. Then for each

Table 9-1 Class/Sub-Class Taxonomy

Class	Pump	
Sub-class	Centrifugal	
Size	3 x 4 x 8	
Shaft Diameter	3.5	Inches
Material of Construction	SS314	
Impeller Diameter	10.5	Inches
Inlet Connection Size	3	Inches
Outlet Connection Size	2.5	Inches
Capacity	250	GPM
Manufacturer	Goulds	
Model Number	3196	
Serial Number	Xyz	

class/sub-class combination, the list of attribute fields you want to record has to be established. The attributes for our Centrifugal Pump were size, shaft diameter, material of construction, impeller diameter, inlet connection size, outlet connection size, capacity in gallons per minute or feet of head, manufacturer, model number, and serial number. The construction of a listing of the class, sub-class and attributes for all of the assets we might expect to populate into an asset management system would be considered creating a taxonomy, or a template or a data entry form. Table 9-1 provides an example of the Pump/ Centrifugal Class/Sub-class taxonomy.

The information in the first and third columns in Table 9-1 is considered the taxonomy (the third column provides the units of measure that are desired for each entry, if applicable). The information in the second column is the actual data that is entered into that template.

Beyond the data entry form, a taxonomy can also include a list of valid choices for each blank field from which to choose your entry. For example, in the Manufacturer field in Table 9-1, suppose we didn't want to allow any entry that was not a valid manufacturer of pumps. We could construct a table of all valid pump manufacturers and then this field could not be filled with free-form text—it could only be filled by choosing a valid entry from the reply table that had been previously created. These kinds of fields are sometimes referred to as validated fields, in that there is a limited choice of valid entries. The reply table that we create for the Manufacturer field would be considered part of the taxonomy. Some reply tables impose text formats (numerical values only, dates only, numerical values with decimal expressions of fractions, etc.).

What about fields in an asset record that are not validated, but can be filled with any characters you choose? These are free-form text

- CLIENT X ! PUMP DESCRIPTION STANDARDIZATION ISSUES
pH PUMP
PH ADJUSTMENT PUMP
BUFFER TRANSFER PUMP
PRODUCT XFER PUMP
CIP RETURN PUMP
RETURN CIP PUMP
CIP CIRC. PUMP
CAUSTIC METER PUMP
CAUSTIC METERING PUMP
PROCESS CHILLED WATER RECIRC PUMP
CHILLED GLYCOL RECIRCULATION PUMP
PUMP, SOFT WATER SUPPLY
PUMP
PUMP
PERISTALTIC PUMP
PERISTALIC PUMP
PUMP ! PERISTALTIC
CIP CIRCLATION ! 2" INLET/1.5" OUTLET
INCO PUMP
CIP PUMP, 3450 rpm
UF702 ROTARY LOBE PUMP
UF!703 INLET PUMP
PUMP, GOULDS 3196
VIKING PUMP AND MOTOR
ROTARY PUMP FOR UF!103A
G&H ROTARY PUMP

Figure 9-5 Description Fields for Centrifugal Pumps

fields. With the exception of a field length limit that might be imposed, no rules are institutionalized in a free-form field (you are in essence free to enter whatever you want). In an asset record, one of the most prevalent free-form text fields is the Description field. This is typically a large field into which you are asked to describe the asset. Figure 9-5 shows a few free-form Description fields for a number of centrifugal pump assets.

Did you notice anything interesting about these descriptions? It should stand out quite prominently that there do not seem to be any formatting rules, or taxonomy, for this field. Each description seems to have been filled with various information, ordered differently, abbreviated differently, etc. The various descriptions were probably entered by different people.

The point here is that a taxonomy can in fact be created for free-form fields as well as for validated fields. The rules of formatting free-form text fields are still necessary even when the software system does not impose formatting rules automatically. A manual or procedural application of those rules will ensure that resulting descriptions are formatted consistently.

In some top-tier asset management systems, functionality exists to allow the description field to be automatically assembled from various attribute data in other fields the Asset record. For example, in such systems you can choose to populate the description field with the data entered into the class, sub-class, and various other attribute fields. The data can be separated by commas, or some other delimiter. In this way you can more readily control what gets entered into these fields. In either case (and whether or not the software allows this), a taxonomy or rule-set is advisable for free-form text fields.

9.8   Ideal Asset Data Repositories

Earlier we talked about various interest groups at a typical plant being interested in and responsible for different silos of asset data. Recalling our example, the instrumentation and controls people keep their asset data in their own systems because of special calibration requirements not typically met by functionality of enterprise asset management systems. Similar separate databases may exist for stationary equipment, fire and safety related equipment, electrical equipment, etc. Is it acceptable practice to allow multiple repositories for various kinds of asset data, or is it advisable to centralize information about all assets in one place?

The answer is that it is considered best practice to have one single repository for all asset data. That repository should be the Enterprise Asset Management (EAM) or Computerized Maintenance Management System (CMMS). Instrumentation technicians may say that the EAM system will not allow performance of the highly specialized and technical calibration activities that are required. Therefore, they must create asset records for the instruments and controls in a separate specialized system, and they would be correct in saying so.

We're not saying that a separate system shouldn't be used for handling the specialized instrument calibration requirements. We are saying that not all of the activities of interest on the instrumentation and controls assets are calibration activities. For example, a failed instrument asset would have to be repaired when failure occurs, or replaced if appropriate. The EAM or CMMS system is the system of record to plan, schedule, control, and document that repair work. Therefore, asset records for the instrument assets must exist in the EAM or CMMS system. The equipment repair history and any spare parts lists would also reside in the EAM or CMMS system with the instrument asset records.

On the surface, this practice would suggest that 1) instrument asset records are required in the EAM or CMMS system, and 2) they are also required in the specialized instrument calibration software program. Basically these suggestions are true. But keeping a duplicate equipment list—at least for the instrumentation and controls assets—in a separate system would require double data entry and data maintenance in two places. The risk of the two lists becoming out of sync is high. As a result, this practice is not considered viable. So are these issues irreconcilable?

The answer is no. The master asset data should be housed in the EAM or CMMS system. The specialized instrument calibration system that needs an asset list should have its asset list populated or fed from the master one residing in the EAM or CMMS. In essence, the duplicate asset list is a slave to the master asset list from the EAM/CMMS.

This strategy of feeding the secondary database from the primary one implies an electronic interface of some kind that would refresh the slave asset list on some frequency, enabling maintenance of the asset master data to be performed in only one place. This strategy of maintaining a master asset list in the EAM or CMMS system makes it easier to adhere to the standard taxonomy rules. The strategy also makes it possible to evaluate criticality, priorities, behavior, history, costs, etc., for all assets in one place.

The instrumentation calibration activity we used in that example is only one example. The vibration analysis technicians would suggest that the rotating machines need to reside in a specialized vibration analysis software system because the EAM system does not do vibration analysis. They would be right to suggest the EAM does not do vibration analysis. But in addition to the vibration analysis activities, many other activities are of interest on the rotating assets. These include spare parts lists, corrective maintenance, work order history, historical costs of maintenance, and run time characteristics. Thus, the asset list required in the vibration analysis software should also be a slave asset list fed from the master asset list in the EAM/CMMS.

This same philosophy applies to all of the specialized require-ments that call for specialized software and asset lists in those software systems. There should be one master asset list, which should reside in the centralized EAM or CMMS system. Any system that needs asset records should be fed electronically from the master asset list in the EAM or CMMS system. The control (or management) of this master data should be performed by a centralized group which has the responsibility for the initial configuration and integrity of the master asset data, and its subsequent upkeep.

It would be a fair criticism of this strategy to suggest that the inter-facing required to feed a number of slave asset lists from a master asset list is an inelegant solution. This solution is one that could be difficult and expensive to maintain, especially given the regular changes that occur in each of the software systems. Because the solution may be difficult is not a valid reason not to do it—the benefits certainly out-weigh the costs inherent in allowing a decentralized, unconnected, and inconsistent treatment of asset master data.

But necessity is the mother of invention. A cottage industry is developing rapidly that attempts to solve this master data management challenge in a more elegant, easy, and simple way. Some companies are developing and already offering various solutions that are designed to connect a wide variety of data from a wide variety of systems and information technology platforms. So the days of relying on internal or contracted information technology resources to do this are numbered. Solutions beyond connectivity solutions are also rapidly developing for data warehousing, data mining and analysis, data synchronization, business intelligence, data visualization and dashboards, predictive analytics, and other interesting processes related to physical assets.

9.9 Enterprise-Level vs. Plant-Level Asset Data Integrity

The argument for central control of asset data integrity at the plant level is well-supported and established as a best practice. Allowing multiple people to create and modify asset master data at will is an outdated and failed strategy. But what about elevating this central control to the enterprise level in an organization that has a number of plants in its portfolio? Is it advisable to centralize the development, deployment, and enforcement of an asset data taxonomy and integrity model? If so, is that more advisable when the plants in the portfolio are similar? Is it also advisable when the plants in the portfolio are more dissimilar?

It turns out that the answer to all of these questions is once again an emphatic yes! Recall from our discussion in Chapter 1 (The Business Case for Data Integrity) the incremental benefits that can be achieved from orchestrating and choreographing the deployment of maintenance strategies (meaning what we do to take care of those assets) for similar assets across the fleet. We explained that the majority of the variation in performance of similar assets is not explained by legitimate differences in geography, vintage, local culture, or skill. The majority of the variation is a result of different, inconsistent, and poor maintenance practices being performed on that equipment from plant to plant.

Figure 9-6 shows two curves that illustrate the variation in maintenance spend and overall equipment effectiveness (OEE) across multiple plants.

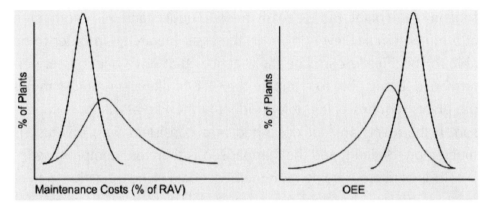

Figure 9-6 Performance Distribution Curves Across a Fleet of Plants

As we can see from Figure 9-6, a few of the plants in this fleet are spending little money on maintenance, a few plants are spending a lot of money on maintenance, and the majority of the plants are in the middle of this classic bell curve. Likewise, a few plants are achieving very good (high) Overall Equipment Effectiveness (OEE), a few plants are posting very poor (low) OEE and most of the plants are somewhere in the middle. Our ultimate goal is to narrow the variation across the fleet and to move all plants to the "good" side of the scale of what we are measuring. In the case of maintenance expense, we want all of our plants to be spending less. In the case of OEE, we want all of our plants to be posting good (high) OEE. Solving for this variation is worth a substantial amount of benefit—financial and otherwise. Being able to solve for it, you first must be able to measure it. We contend that without a centralized and standard asset data taxonomy, it is impossible to measure this variation, let alone solve for it.

Although it is true that many companies do in fact measure and compare OEE across multiple plants, this is typically done at a much higher level of granularity—for example, the entire plant or the entire section of the plant. But we really need to understand this information at a more granular level (lower in the asset hierarchy) in order to be able to specifically correlate the practices that drive better asset performance. Being able to compare the OEE of like assets across multiple plants requires being able to identify the like assets for comparison. If the assets are not described in a consistent way, this task is much more difficult, and the comparison is therefore compromised.

To illustrate our contention that asset data standardization is necessary to measure variance across multiple assets, here is a more specific example. Suppose we want to know the Mean Time Between Failures (MTBF) for the asset class and sub-class "Centrifugal Compressors" installed in similar service across all of our plants. We must first be able to identify and find all of the like occurrences of this asset class/sub-class. If we allowed each of our plants to describe their assets as they saw fit, it is undeniable that there would be a wide variety of ways that centrifugal compressors were named and described from plant to plant. In fact, the reality is that, at any single plant, it is highly likely that centrifugal compressors are named in different and non-standard ways even within a single plant!

In this case, our goal is to find all of the centrifugal compressors in the enterprise, compare their MTBFs (basically building the normal distribution curve for them), then further investigate why there are varying performance characteristics on similar assets (undoubtedly finding a variation in the maintenance activities that are being performed on them), and ultimately solve for that variation (moving all of the compressors to the good side of the MTBF curve). To achieve this goal, it is an absolutely vital and essential prerequisite that a standard taxonomy for the master asset data be created, deployed, and enforced across the enterprise.

There is resistance in some company cultures to impose central mandatory standards on plants that have traditionally enjoyed considerable autonomy. The autonomy of the plant manager to deal with the unique local environment is considered important in many companies. We're not arguing for ending that autonomy. Most companies believe that mandatory requirements from a corporate central group are not well-received. Furthermore, most companies understand that simply willing something into being from the corporate level doesn't make it so. Human culture dictates that people accept and support new ways of doing things if the individual people want to do it that way. The traditional approach has been to create central advisory groups that can suggest (but not mandate) certain standards. These groups have also tried to help the plant's employees conclude that a new way of doing something is a good thing and then eventually enjoy the results.

This approach is fatally flawed in several ways. First and foremost, it is very rare today to find a consistent and standard pattern across multiple plants, even in companies that have had these "centers of excellence" advisory groups when it comes to asset master data. So number one, the approach doesn't work. Second, suppose you argue that bringing people around to a certain way of thinking or doing—if done well—will eventually result in better buy-in and higher levels of compliance and consistency (and we already said we don't think that works). It certainly takes a lot of time to get there.

Finally, a standard and consistent asset master data taxonomy is an absolutely fundamental building block of world-class asset management performance, and resultant business performance. Having one should not be negotiable. How could we not insist on that, unless we were willing to forfeit a considerable entitlement of company performance?

Asset data integrity standards should be likened to the financial accounting chart of cost centers and accounts in the company's general ledger, or to the standard way that safety statistics (like injuries) are categorized, recorded, and reported. It is unthinkable to allow a plant manager to decide how to account financially for the plant's expenses. How would the company ever be able to roll up all of the expenses from around the globe and prepare the required financial reports? It is unthinkable to allow an individual plant to decide what is considered a recordable injury.

Likewise, it should be unthinkable to allow an individual plant to decide how to format and manage the integrity of its asset master data! In any industrial company, the company's ability to achieve its strategic goals is contingent on the performance of its physical assets, often the most valuable asset on the company's balance sheet. Why would we leave that up to individual plants and people to decide how to best manage such a critical and valuable resource in our company? Having a standard and consistent asset master data integrity model—one that is mandated, enforced, non-negotiable, and considered strategic to the company's ability to perform—is as fundamental as the chart of accounts!

Chapter 10

Building an Enterprise-level Inventory Catalog Data Integrity Model

> Critical spares are great to have when you need them. The important thing is to be able to find them and know that they will fit.
>
> *Focus Group*

10.1 The Model For Material

In the previous chapter, we discussed the enterprise-level data integrity model for physical assets installed and operating to perform a particular function. This chapter will discuss the equal importance of an enterprise-level data integrity model for the material items that are not installed and operating, but are considered spare parts kept in stock, or ordered when needed. These items are usually documented in some kind of materials management software application that includes a data file for inventory catalog records.

To be clear, there are several categories of items that could be represented in the catalog data file of a materials management software application. Primary categories are Raw Materials, Finished Goods, and what we call MRO commodities (Maintenance, Repair, and Operating supplies). It is the MRO category with which this chapter will concern itself, though many of the concepts discussed here apply very readily to the other major categories of items.

When it comes to managing materials in the MRO category, the software applications that should be used to control these inventories are the same applications used to manage maintenance and assets. These include Computerized Maintenance Management Systems (CMMS), Enterprise Asset Management (EAM) systems, and inventory management modules of Enterprise Resource Planning (ERP) systems (e.g., SAP). These systems provide the ability to create a catalog of items in the MRO category. Each item is typically represented by some sort of an inventory master record in the master catalog data file, with each inventory master record consisting of myriad information describing the item. These records usually are identified with a unique identification number, one or more descriptions, keywords and parametric fields used to search for and find the part in a large catalog file (often containing thousands or even hundreds of thousands of items), storage locations including row and bin numbers, quantities on hand, reorder points, reorder quantities, supplier information, etc.

Traditionally, inventory catalogs have been created at a plant level, not at an enterprise level. Over a long period of time, many people enter new items. For example, new parts are acquired for storage, new equipment requiring new spare parts are installed in the facility, and vendors change their order part numbers. An objective system is needed for governing naming and numbering conventions, description formatting, spelling, and configuration of information. Otherwise, inconsistent ways of describing these items will result in many inefficiencies and wasted effort. As described in a previous chapter, each maintenance worker wastes approximately 30–45 minutes per day looking for spare parts either in a poorly formatted catalog or physically in a storeroom. This is time that can be saved with a properly and consistently formatted and maintained items catalog. Significant gains in worker productivity are available by investing in an enterprise-level inventory data taxonomy.

Beyond the worker productivity issue, approximately half of all maintenance dollars spent annually in the developed world are spent on spare parts vs. labor. In geographies with lower cost labor, the percentage of total maintenance spent related to spare parts is significantly higher than 50%. There are usually more inventory catalog records just in the MRO category than there are physical assets, so it is arguably more important to bring discipline to this area of master data than even to the all-important physical asset area. For obvious reasons, we can't leave this large part of our business to haphazard and disparate master data strategies.

10.2 What Is a Spare Part?

When we are talking about spare parts data, just what are we talking about? Recall in our previous chapter, we described physical assets. We discussed that at some point an object ceases to be considered a physical asset—primarily when that object is not a repairable or maintainable item, but instead one that is thrown away and replaced when it fails. Our examples included an automobile engine that could be considered an asset. But then we discussed how that engine is broken down into many systems. Each system can be broken down into many components, and eventually you reach a level of granularity where the component is not an item that is repaired. Using that same example, the many nuts and bolts and other fasteners that are used in that engine are certainly not maintainable assets, but are considered spare parts.

Terminology used to refer to spare parts and other categories of stocked or ordered materials can be confusing at times. Terms like *spare parts, materials, supplies, items, goods, inventory, commodities*, and others are sometimes used interchangeably to refer generally to stocked or ordered objects. Some of these terms, technically speaking from an inventory management standpoint, do have specific connotations. For purposes of this chapter, however, we will use the term *items* to refer to the MRO objects in the "items" catalog.

10.3 Items Classification

MRO Items can be broken down into a kind of hierarchy of classifications. Generally in the asset management world, there is usually a division between stocked and non-stocked catalog records for items. As the name implies, stocked items are physically kept in stock in a storage facility, whereas non-stock items are ordered when needed. Modern materials and asset management systems encourage both stocked and non-stocked items to be cataloged.

Another primary breakdown distinguishes between spare parts items and generic supplies items. Generally speaking, spare parts items are those that are asset make/model-specific; they are usually available only from the original equipment manufacturer (OEM) who manufactured the asset. (In some cases parts that were previously available only from the OEM have been reverse-engineered by other manufactures and in these cases the parts can be sourced from suppliers other than the OEM.) Generic supplies, on the other hand, are generally items that are not OEM-specific; they can be acquired from a number of sources and, in many cases, can be used interchangeably on many different assets. Examples of generic items include fasteners such as nuts, bolts, screws, and washers. Other examples of generic items include pipe fittings such as elbows, tees, reducers, and flanges.

Some commodities are confusing, claiming to be asset-specific but in actuality being generic at the same time. One example of this is bearings. There are different types of bearings used in various rotating machines. OEMs sometimes assign their own part number to bearings in their assets and want you to buy your spare bearings from them. In actuality, almost all bearings are manufactured generically—in fact, most OEMs buy their bearings from a generic manufacturer. The generic bearing manufacturers make their bearings in various sizes and configurations for various applications and duties. You can buy a bearing with identical dimensions, materials of construction, configuration, etc., from more than one manufacturer. It is often less expensive to buy your spare bearings from one of the generic manufacturers than from the OEM, though this is not always the case. Other such commodities include seals.

For commodities like bearings and seals, and others, there exist interchange manuals (hard-copy or electronic) that provide the equivalent part numbers for the various bearing manufacturers. So, when you remove a bearing from a pump or fan and need to buy a new one, if you can identify the manufacturer and part number of that bearing, you can look up in the interchange manual all of the equivalent bearings with their manufacturers and part numbers. The interchange gives you more flexibility to buy your spare bearings from a variety of sources. Consolidating or standardizing on one or two manufacturers of bearings, seals, etc., can often save significant cost. It can also serve to significantly reduce inventories, saving not only money with large quantity discounts, but storage space as well.

Commodity codes are often the next level of classification of items. Examples of commodities include:

- ❏ Power transmission parts
- ❏ Pipes, valves, and fittings
- ❏ Fasteners
- ❏ Lumber
- ❏ Metal stock
- ❏ Electrical
- ❏ Electronic
- ❏ Computer supplies
- ❏ Chemicals
- ❏ Cleaning supplies
- ❏ Personal Protective Equipment (PPE)

There are many more.

Most software application catalog item master records allow the assignment of a Commodity Code. This code is usually chosen from a reference table previously created, allowing the Commodity Code for a specific item to be chosen from a pull-down list of valid choices. Creating a sound commodity code structure is an important foundational step in any good items taxonomy.

Within each commodity, the next level in a sound classification system for items would be the Class—also referred to in some systems as a noun. Each Class has associated with it multiple Sub-classes, often referred to as modifiers or adjectives.

Like the assets in the earlier chapter, each item can be classified by item Class and Sub-class at the item level itself. The following are a few examples of item Classes and Sub-classes:

❏ Item Class = Bearing; Item Sub-classes = Ball, Pillow Block, etc.

❏ Item Class = Elbow;   Item Sub-classes = Pipe, Tubing, Conduit

Like with the assets, Item Class and Sub-class are important in that they are used by most inventory information software systems to define various attributes of the database record that describes an item. For example, in our bearing ball example, the following may be the list of attributes for any record that is assigned the Class "Bearing" and the Sub-class "Ball." (Note that the italicized entries are sample data that may be filled into any of the attribute fields for a particular record. However, the attribute field names remain constant for all item records that are assigned the Class/Sub-class of "Bearing, Ball.")

❏ Manufacturer/Vendor Name

- *SKF BEARING*

❏ MFG/Vendor Part Number

- *60032RS*

❏ Alternate MFG(s)

- *FAFNIR BEARING*

❏ Alternate Part Number(s)

- *9103PP*

❏ Attribute / Specification Template

- ROW: *SINGLE*

- THRUST: *RADIAL*

- SERIES: *EXTRA LIGHT*

- SEAL: *2 CONTACT*

- I.D.: *17 MM (0.6693 IN)*

- O.D.: *35 MM (1.3780 IN)*
- WIDTH: *10 MM (0.3937 IN)*

In most software applications that have item catalogs, at least one description field is usually used as the description of the item in terms familiar to the item's user (in our case, typically the maintenance worker). In some cases, there is more than one description field. Additional description fields may include a Search Description and a Purchasing Description (in terms specific to the suppliers vs. the users of the item). Sometimes there is a Long Description and a Short Description. It is important to recognize the particular functionality of your system to devise a sound strategy for creating data in these fields.

Many systems allow free-form text entry into Description fields. As you can imagine, this practice causes inconsistent formatting over time, with different people entering data into these free-form fields at will. Examples abound where Description fields are formatted with abbreviations, different ordering of terms, different delimiters separating discreet data elements, etc. For example, sometimes the word Bearing is abbreviated and not always the same way (e.g., BRG, Brg). Or, numeric values may be entered sometimes as fractions (1/2) and sometimes as decimal expressions (0.5).

All of this variation makes searching for items very difficult and sometimes impossible. A formal and rigorous taxonomy or rule-set is required to ensure consistent and usable descriptions. Some consultants specialize in classification of materials items and they bring extensive pre-developed taxonomies and rule-sets that can help considerably in preparing standardized material catalogs.

Modern software applications for item catalog management contain functionality that prohibits entry of free-form data into description fields. Instead, they provide for the population of the Description fields by assembling automatically, into a pre-defined order, attribute data from other discreet fields in the record.

Using this method, notice the following system-generated description field using the attribute data from the template previously shown:

❏ Material Description (a standardized stringing together of all the individual static data elements)

❏ Bearing, Ball, Single Row, Radial Thrust, Extra Light Series, 2 Contact Seals, 17 MM (0.6693 IN) I.D., 35 MM (1.3780 IN) O.D., 10 MM (0.3937 IN) Width, SKF Bearing P/N 60032RS, Fafnir Bearing P/N 9103PP

Even if your software does not have such functionality and allows free-form text entry, there must be a formal and strictly enforced taxonomy that is used to guide the entry of data into these fields.

There are specialty consulting companies that provide services and software specifically designed to normalize item catalog data. They usually bring with them pre-developed taxonomies with Class and Sub-class structures, along with appropriate attribute tables and extensive valid reply tables for each attribute field. These pre-developed taxonomies are significant accelerators; they should always be considered over developing one from scratch. These taxonomies are also often already cross-referenced with national and other standards including UNSPSC, which is an attempt at a global standard numbering and naming system for all goods and services.

10.4 Static Data vs. Dynamic Data

As you can see in the Description example above, the information (or data) that is displayed is mostly descriptive information about the item. As with Assets, information describing the physical characteristics about the item that tends not to change over time is considered static data. Conversely, information that has the likelihood of changing frequently is referred to as dynamic data. In most cases, the static data is created once; it is then appropriate (accurate and complete) for long periods of time. Static data is also known as Master Data.

Information about an item that has the tendency to change over time—dynamic data—is either populated automatically through the course of transactions (like work orders, inventory issues, inventory returns, inventory receiving and binning, and inventory purchases) or the dynamic data has to be populated manually when changes occur. Examples of fields of dynamic data in a typical item master record that would tend to change frequently are: quantities on hand, last purchase price, average unit price, quantity reserved, quantity on order, usage history, order history, and issue history.

10.5 Ideal Item Data Repositories

Earlier we talked about various categories of items that can be managed by materials management software applications. These include the primary categories of Raw Materials, Finished Goods, and our category of interest MRO, or Maintenance, Repair and Operating commodities. The ideal repository for the MRO commodity items is the system of record used for asset management and maintenance. Keeping the MRO commodity items in the asset management system is critically important because there is an intimate relationship between the MRO items and the physical assets. Separating these two large pieces of master data is a mistake. Again the system in which to keep the MRO items is the Computerized Maintenance Management System (CMMS), the Enterprise Asset Management (EAM) system, or the materials management module of the Enterprise Resources Planning (ERP) system—whichever is being used for asset management and maintenance.

10.6 Enterprise-Level vs. Plant-Level Item Data Integrity

The argument for central control of item data integrity at the plant level is well-supported and established as a best practice. Allowing multiple people to create and modify item master data, with no established rules, is an outdated and failed strategy. But what about elevating this central control to the enterprise level in an enterprise that has a number of plants in its portfolio? Is it advisable to centralize the development, deployment, and enforcement of an item data taxonomy and integrity model? If so, is that more advisable when the plants in the portfolio are similar? Is it also advisable when the plants in the portfolio are more dissimilar?

It turns out that the answer to all of these questions, as with assets, is yes. Why? Recall from our earlier discussion in this chapter the amount of time wasted when items are not described in a way that can support efficient searching and finding of spare parts needed for repairs. This benefit admittedly is a plant-level benefit. However, extending this policy to the enterprise level allows for efficiency in that there is a need for the creation and enforcement of only one standard vs. many.

But incremental advantages that can be gained by elevating standards for items classification to the enterprise level are significant.

Consider the earlier argument that leveraging purchasing power across the enterprise, reducing inventories safely, sharing inventories appropriately across facilities, standardizing on items for company-wide use, measuring inventory spend with certain suppliers, and many other prudent business practices are severely hindered without a sound, standard consistent strategy for naming and managing items.

Modern enterprise-level materials management systems recognize the value of enterprise-level standards for a corporation. Therefore, they provide integrated functionality to address currency and language, and even alphabet character-set differences that exist with global companies. Don't be deterred by the seemingly daunting task of a global standard for items taxonomy. The task is not simple, but the benefits are orders of magnitude greater if an enterprise-level of items data integrity is developed and implemented.

Chapter 11

Data Integrity Assessment

> The good news about computers is that they do what you tell them to do.
> The bad news is that they do what you tell them to do.
>
> *Ted Nelson (Information Technology Pioneer)*

11.1 Data Quality Dimensions—The Beginning

Data assessments are used to determine if organizational data can be used as reliable information; the Business Intelligence (BI) of every well-managed organization. At its foundation, data must adhere to rigorous standards of compliance and sustainability if an organization is serious about good data quality. These compliance and sustainability measurements are referred to as DQDs, or Data Quality Dimensions. Each dimension listed in Table 11-1 is a separate DQD; the organization needs to assign a benchmark measurement for each. Having these benchmarks allows the assessment effort to determine gaps in the quality of asset data present. DQDs may be objective or subjective, depending on whether they are based on generally accepted best practices (or quality benchmarks) for a given dimension or on the experiences of the organization using them. Whether objective or sub-jective, DQDs should be used to baseline an organization's data qual-ity and ultimately determine the effort required to correct data-related problems.

Looking at the DQDs in Table 11-1, you might logically ask, "How do I assign measures to these elements? There are two approaches, objective and subjective.

❏ **Objective DQDs.** Some elements are objective such as complete-ness, free-of-error, and consistent representation. These can be asso-ciated with industry best practices. The metrics can be provided by the specialty consultants that work in this arena.

❏ **Subjective DQDs.** The metrics associated with subjective elements are determined by organizational experience and specific goals. Measures for items such as accessibility, believability, and under-standability are based on perception. This information can be obtained from the organization through surveys, personal experi-ence, and open discussion.

The organization should continually refine and improve on each of the data quality measurements to further increase the value of its data. As part of the sustainability plan, the data should periodically be monitored and audited against the DQDs to assure its continuing quality.

Missing any of these dimensions in existing asset data manage-ment systems can cause an individual user or an entire organization to mistrust the data. This problem can then result in a significant loss in productivity and Business Intelligence (BI). Often companies expect that upgrades to software in which asset data is resident will correct these poor data quality issues. Technology or software upgrade initia-tives are always accompanied by an IT-driven data transfer activity, but

Table 11-1 Data Quality Dimensions

Dimensions	Definitions
Accessibility	The extent to which data is available or easily and quickly retrievable
Appropriate Amount of Data	The extent to which the volume of data is aligned to the level of detail required for the task at hand
Believability	The extent to which data is regarded as true and credible
Completeness	The extent to which data is not missing and is of sufficient breadth and depth for the task at hand
Concise Representation	The extent to which data is compactly represented
Consistent Representation	The extent to which data is presented in the same format
Ease of Manipulation	The extent to which data can be readily modified to be applied to different tasks
Free-of-Error	The extent to which data is correct and reliable
Interpretability	The extent to which data is clearly defined and is in appropriate languages, symbols, and units
Objectivity	The extent to which data is unbiased, unprejudiced, and impartial
Relevancy	The extent to which data is applicable and helpful for the task at hand
Reputation	The extent to which data is highly regarded in terms of its source or content
Security	The extent to which access to data is restricted appropriately to maintain its security
Timeliness	The extent to which the data is sufficiently up-to-date for the task at hand
Understandability	The extent to which data is easy to comprehend
Value-Added	The extent to which data is beneficial and provides advantages from its use

Pipino, L. L., Lee, Y. W., & Wang, R. Y. (2002). Data Quality Assessment. Communications of the ACM , 211-218.

very often the integrity of the legacy data is not seriously assessed. In this common scenario, bad data in the old systems are transferred into the new system without enhancement. Many companies have learned the hard way that simply upgrading to the latest Enterprise Asset Management (EAM) or Computerized Maintenance Management System (CMMS) does not create better BI. In fact, many organizations overlook or undervalue the benefits of having reliable data as a part of this process.

The data assessment should be a prerequisite in all system integration, merger and acquisition projects, and consolidation of disparate systems that require enterprise-wide single system integration. Data analysis should also be required for organizations that have already consolidated master data to an enterprise-wide single system, but whose data is recognized as being corrupt and not standardized. Even in companies that are not changing systems and will continue to use existing systems, the data must be reliable, making an assessment a worthwhile effort. This enables the business processes using that data and the various systems to generate good fact-based decisions and better business process performance.

In many cases today, companies are standardizing their technology platforms and systems architecture across the enterprise. As part of this trend, companies are retiring many legacy systems, then retrieving the legacy data for loading into new standard systems. Bringing all the data into an enterprise-wide single system is known as Enterprise Information Integration or EII, a process of information integration, EII provides a single interface for viewing all the data within an organization, and a single set of structures and naming conventions to represent this data. The goal of EII is to get a large set of heterogeneous data sources to appear to a user or system as a single, homogeneous data source.

Your scenario may involve moving legacy data from multiple systems into a new central system or you may intend to keep your existing system. In either case, moving data without cleaning it, or assuming that what you currently have in place is good enough, will not address the problem of data integrity.

The real solution is to have your data in a condition that has attained acceptable benchmark-quality levels for the various DQDs. Attaining those levels of quality creates confidence in the data and the system, enabling the users of the data to make informed judgments. Reliable data also promotes a culture for sustaining the integrity of the data—high quality data make the user's work easier so there is an incentive to keep the data quality high.

11.2 The Approach to the Assessment

To determine the integrity of an organization's data and the consequences of not having dependable data, the organization needs to apply cross-functional knowledge, tools, and a commitment to the assessment effort. Unfortunately, many organizations do not have the expertise resident in the company to efficiently assess their legacy data. They often rely on the System Integrators (SI) to solve the data problem. System Integrators are typically third-party companies that specialize in the information technology aspects of computer application implementations. However, these third parties often don't have the subject matter expertise to recognize inaccurate, incomplete, or erroneous data.

Some companies have the subject matter expertise resident internally, but the people with that knowledge and expertise are usually tasked with other primary duties that diminish their ability to invest the time and concentration necessary to assess data integrity.

Other organizations simply do not understand the lack of data integrity in their legacy systems.

In all of the above cases, contracting with a consulting firm that specializes in asset data assessments and improvements is a worthwhile investment that can yield significant returns for an organization.

11.3 The Initial Steps

Quite often during a project to improve the data in the company's existing system, someone in the organization recognizes that the data is not reliable in its current state. Sometimes when existing systems are going to be replaced, someone recognizes that simply moving data from the legacy system to a new system will not achieve the desired outcome of having good Business Intelligence. This realization often happens when the organization is already months into the project, or even after such a project is completed. If the team recognizes the data deficiencies mid-stream and decides to do something about it, the project team needs to re-align resources, budgets, and timelines to adjust for the newly discovered importance of obtaining reliable data. This isn't the way it is supposed to happen.

In a more ideal scenario, the project team will conduct a thorough assessment of the existing data during the planning phase of the effort. Most often, an assessment of this nature reveals data problems that require correction. Having this information upfront allows the project team to develop a cost–benefit analysis, a realistic budget, and resource requirements. The team can then incorporate this knowledge into the project plan. A proactive data assessment will result in a project plan that addresses the data integrity problems. A robust data assessment and enhancement process will typically include the following:

❏ Legacy Data Assessment—The legacy data is evaluated against the desired benchmark quality levels for relevant DQDs, and data enhancement activities and rules for data transformation are established.

❏ Data Standardization—The application of the data enhancement rules will standardize the format of the legacy data (whether a new software application is being introduced or the legacy data will be extracted, enhanced in a specialty data enhancement system, and returned to the original system).

❏ Data Enhancement—Where legacy data is not available for fields in the to-be data model, the gaps are filled or resolved.

❏ Data Relationships—Where transformed data refers to other data tables (e.g., the definition of codes are housed in a separate data table or system), the data is linked to other data sources.

❏ Data Sustainability—The creation of a policy dictating how the data will be sustained throughout the enterprise and over time.

❏ Data Quality Reporting—The creation of a system of monitoring data quality metrics for ongoing data quality compliance.

11.4 The Assessment—General Comments

At its core, a data assessment measures the various data points against an acceptable quality standard or benchmark with the appropriate tools and knowledge to make the comparison. (Chapter 12 will address standards and benchmarks for asset data and Chapter 13 will address standards and benchmarks for data related to material items.) To determine what good data looks like, the data quality dimensions from Table 11-1 must be defined and applied to the overall data assessment.

In addition to identifying gaps in legacy data, the data assessment by its very nature will expose deficiencies in both the work management system and the policies of the organization that allowed the data to become unreliable in the first place. As discussed earlier, it is often difficult for an organization to resolve data integrity problems with internal resources. A good specialty consultant firm will objectively inform the organization of the data deficiencies and recommend initial steps to close the data gaps, The firm will also recommend measures and policies that, over time, will sustain the data in compliance with the rules.

Cultural indifference to the value of reliable asset data often creates the conditions that lead to the data reliability problems. Given that some changes in policies may be needed, the organization should assess its readiness for change during the consideration of a comprehensive data integrity assessment. Indifference to or ignorance of the value of data integrity can happen at various levels of the organization. Resistance to change at any level, if not identified and addressed through a formal change management program, can lead to systemic and adverse affects on the data. Focusing only on the technical challenge and ignoring the cultural challenges will lead to sub-optimal results. These will include less-than-optimum data to start with and policies that will allow the data to degrade again.

11.5 The Assessment Process

Step 1: Pre-Work—Obtaining Support

Software and consulting companies specializing in data assessments exist. Although many of these companies address data related to customer information (companies, names, addresses, etc.), a few focus on physical asset data. For assessing enterprise asset data, it is these latter companies that provide the greatest value; they bring the subject matter expertise, along with the pre-developed content and taxonomies specifically related to physical assets. One of the authors of this book is the founder of one such company, Management Resources Group, Inc. Utilizing the existing data assessment software applications (usually available from a consultant specializing in this type of work) makes handling of the typical large volumes of data associated with an assessment a manageable task.

Step 2: The Beginning

As a starting point, it is advisable to obtain all the Master Data tables and files (those that relate to asset and material items) and all data dictionary elements (those that describe the asset) contained in the files and tables. Extraction of the specific asset data will occur in Step 5. Note that Master data is distinct from Transactional data. *Master* data provides descriptive information about the physical asset or material item. *Transactional* data provides information about the various activities related to an asset or material item, such as work order history or purchase history. We are going to focus on master data for our assessment. Master data typically resides in a subset of data tables in a software system.

Although fairly comprehensive data assessments can be accomplished with only a representative subset of the Master Records themselves, you should obtain all of the data dictionaries for the master data tables. This will give you the ability to understand the table names and the column names for the various data elements that reside within the tables. Think of a data dictionary as the first row in a spreadsheet that has the column names prompting the viewer to know what kind of data is being requested in each column. Think of the data entered into the rows and columns as the master data itself. In this spreadsheet analogy, the first row with the column names is the data dictionary. The information entered in the rows and columns is the master data, with each row representing one physical asset or one material item. Having all of the data dictionaries for all of the master data tables decreases the probability of an important data quality dimension being miscalculated.

Once the data assessment software tool is selected—as well as a specialty consultant, if appropriate—and all of the master data tables and data dictionaries are retrieved, the assessment can begin. Review the various legacy master data tables received to determine if the complete set that was requested from the functional departments in the organization is actually what was received. Incomplete tables or truncated data elements are very common problems. The people or department responsible for the data extractions typically have abundant IT and data extraction capabilities. However, they are usually unfamiliar with the data content and would not be able to recognize gaps or problems.

To avoid an incomplete legacy data set, obtain a document that lists all the tables, by table name, and the columns of data in each table, from the legacy system(s). The documentation that lists the tables and columns in a legacy software system is often referred to as a *data file structure* (or a *data dictionary* as we discussed earlier). The IT people will understand this terminology. With this information you can determine which tables are needed to conduct the assessment.

Step 3: Clarification

Depending on the format of the data dictionary documents, it may be necessary to do some more work to determine the meaning of some or all column names. Sometimes the column names are not indicative or intuitive enough to determine what kind of information is housed in a particular column. Sometimes the column names are cryptic or non-intelligent. In these cases, it may be necessary to view samples of the data directly in the software application through its screen user interface. The goal is to create an understandable description for the column name, clearly defining what kind of data is meant to be stored in each column of each table. This detective work will help you develop a complete picture of where the data of interest is actually meant to be stored in any legacy systems.

The column names or the short column descriptions of what kind of data is meant to be stored in each column is referred to as the metadata or meta-information. The *metadata* is the *data about the data*. It is important to understand because it is foundational to the entire data assessment process. If, for example, you consider data about a 60-watt light bulb in an inventory catalog, the metadata might include the class (bulb), sub-class (light), wattage (60), color (soft white), base style (candelabra), frosting (clear), etc. You can view the data in the record, but without knowing the metadata, you may not be able to recognize the significance of each data element. For example, if you saw a data element of 60, how would you know that was meant to represent the wattage without knowing the metadata?

Metadata organized into a structure is sometimes referred to as a taxonomy. Some of those specialty consultants already have extensive metadata taxonomies for asset-related and material item-related data.

These taxonomies provide a vastly advanced starting point for a data assessment and standardization effort.

Step 4: Review

Next, review and gain a full understanding of the data dictionary documents, including the understandable column names that may have resulted from the detective work mentioned above, to determine which tables and columns are of interest for the data assessment. Some of the tables will contain master data, like descriptive information about physical assets or material items. Some tables will contain transactional data, like work order history, purchase order activity, etc.

Step 5: Extraction

Once your understanding of the data structures is complete and clear, you need to determine which data should be extracted for assessment. There are two approaches. The first is to stress to the IT people which tables of data are needed and request <u>all</u> the columns and <u>all</u> the rows in each of those tables. Even if some of the data will not be useful for the purpose of the data assessment, this approach minimizes the chance of missing or leaving behind what may be very relevant data.

Alternatively, you can have the appropriately skilled people (those who recognize the content) carefully identify the tables, columns, and rows of interest ahead of the extraction; then allow the extractors to provide only that data. In either case, the appropriately-skilled people are the ones who should be determining which data should be extracted and whether the full set of data has in fact been received. These approaches do not guarantee a successful and complete extract the first time, but they do lessen the need for and frequency of subsequent requests for additional information.

Step 6: Analysis

With the documented table structures completed, including the understandable column names in each table, the task of reviewing the contents of each table (the actual data in the rows) can begin.

Although there are many methodologies that can guide the actual assessment process, a typical early activity is to establish an understanding of how the various columns in each table were actually used, regardless of how they were intended to be used. For example, there may be a column in an asset master table whose metadata (column name) is *Model Number.* Only a subject matter expert could recognize the significance of the data and be able to determine if all of the entries in this column are in fact model numbers. Some data may instead be something other than model number. There may have been a good reason when the original system was implemented to establish a use for this column other than what was originally intended. It is very typical to find this case in industry.

Complicating this common reality, the alternate use of some columns is not necessarily consistent from row to row in a particular table. For example, the Model Number column in the asset master table may have been used to house Serial Number for some assets, Alternate Asset Number for other assets, and Model Numbers (as originally designed) for yet other assets. All of these issues relate to the data quality dimensions of *Free of Error* (the extent to which the data is correct and reliable) and *Relevancy* (the extent to which the data is applicable and helpful for the task at hand). The new system may in fact have a column for Serial Number. But unless the reviewer recognizes that the Model Number column in some rows of that table houses a Serial Number, that information will be misdirected. As a result, fields needed in the new system may be unnecessarily left blank. The assumption may have been that the Serial Number was not available in the legacy system, but in reality it is available. It is just not where one would expect it to be.

Another good example of the need for subject matter expertise deals with requirements for data in a new system that may appear to be missing in a legacy system. Suppose a new asset management system required a piece of data called Class in one column, referring to the class, or the noun name, of an asset. Suppose further that the legacy system has no such column or data. However, suppose the legacy system has an asset number that has some intelligence, including an alpha code that signifies a class of asset. An asset number of P-101A might not appear significant to a casual observer, but it could readily be established that the P indicates this asset is a Pump, and therefore we can derive the data for the new Class column from the legacy asset number—in this case, Pump. Once again, without the subject matter expertise, or automated features to recognize these kinds of patterns, it is likely that a simple IT-driven data extraction and load process would miss this data and render columns in the new system blank.

Other typical data quality dimensions on which the review process for asset data should concentrate include *Completeness* (fill-rates—the extent to which the data is not missing), *Consistent Representation* (the extent to which the data is presented in the same format), Believability (the extent to which the data is regarded as true and credible), *Interpretability* (the extent to which the data is clearly defined and is in appropriate languages, symbols, and units) and *Timeliness* (the extent to which the data is sufficiently up-to-date).

The extent to which various data quality dimensions are analyzed and their relative importance can vary by situation. It is always advisable to establish the scope (which data quality dimensions should be included in the assessment) and relative importance or priority of each data quality dimension. The scope must be established in the context of the to-be model that is desired. If achieving best practices in a business process is the desired to-be model, then it is likely that more work will be required to bring the legacy data into a fit-for-purpose condition. If, on the other hand, the to-be model is simply to replace legacy technology and not advance business practices, then it is likely that less work will be necessary. Regardless of how advanced the desired to-be state is, however, our experience is that it is almost always insufficient to simply map and convert legacy data into a new system without a good data assessment and some enhancement work.

Step 7: Report and Action Items

At the conclusion of the assessment, an initial report should be created outlining the understanding of what data is currently in hand, its level of quality with reference to the DQDs, and the extent to which the data conforms to the format requirements in the to-be data model. The report should detail the activities that will be necessary to render the legacy data into a fit-for-purpose condition. It should also prescribe policies that should be enacted to prevent the data from degrading from the applied standards in the future. Remember, all of this applies, regardless of whether the legacy data is being converted into a new system or intended to be improved and returned to the legacy system.

11.6 Moving Forward

In this chapter, you can begin to appreciate the complexity associated with data integrity assessment and perhaps why specific software applications have been created to deal with this complexity.

Simple spreadsheets or database applications can't possibly provide the facility and functionality necessary to perform a comprehensive data assessment in an efficient and cost-effective manner. It is almost always more than cost-justified to engage an appropriately qualified specialty consultant, along with their software and taxonomies, to assess your data. (It should be noted that these firms and tools are also usually very well-equipped to efficiently correct and enhance your data once gaps are identified.)

In the next two chapters, we will address at a relatively high level some of the standards and benchmarks that indicate acceptable levels of data quality when moving from an existing set of legacy systems to a consolidated or enterprise level software application. Chapter 12 will address physical asset data attributes, and Chapter 13 will address materials items data attributes. Chapter 14 will focus on the peculiarities of data integrity clean-up efforts when the data in the legacy system will be re-loaded back into the same system—as opposed to a new system.

Chapter 12

Assessment Details— Assets and Material Items

> Excellent asset master data is not optional;
> it is foundational to everything
> an industrial company does.
>
> *Data Integrity Work Team*

12.1 Similar But Different

An assessment has two major areas of focus—assets and the material items that support the repair of the assets. There are similarities when assessing the quality and needed repair for both types of data as well as differences that must be taken into account if you wish the outcome of the assessment to be successful.

Most organizations undertake full-blown data assessments when they are moving their data from legacy systems into a newly purchased application. The reason—the users of the data recognize the problems with the existing data and use the occurrence of the installation of the new computer system to fix the data problem. With this in mind, the balance of this chapter will be dedicated to a data clean-up effort when a new system is being implemented. Data clean-up efforts undertaken when the legacy data is going to be put back into the same system are somewhat different. For these kinds of data clean-up efforts, we refer you to Chapter 13 which describes two approaches—*Big Band and Fix It As You Go*—to address clean-up of existing system data.

12.2 Assessing Asset Data

Having accurate, complete, and consistent asset master information is the cornerstone to any asset management strategy. The ability to track physical assets from the time of acquisition until retirement is crucial for life cycle analysis, production, maintenance strategies, and accurate financial reporting in business. In addition, asset master data records are the foundation of any EAM or CMMS. When asset records are incomplete or suspect, then transactional data necessary to drive business decisions are also incomplete or suspect. Asset master records consist of asset or equipment data—manufacturer, model, serial number, equipment specifications, bills of materials (BOMs), locations (physical and relational), coding information, cost centers, and documentation references.

As we have discussed all of these data elements are considered static data in that they don't change frequently. Another type of data is referred to as transactional data. This kind of data does indeed change very frequently. Master Data Management concerns itself with the static data elements.

Asset master data supports all subsequent transactional data related to assets. Transactional data based on faulty master data is misleading at best, and is worthless and costly at worst.

Typical frustrations expressed regarding asset data include:

❏ Assets that exist in the plant but are not represented in the master asset file of the EAM or CMMS

❏ Multiple systems housing conflicting information about the same assets

❏ Inability to find assets in the EAM or CMMS because of nonsensical or inconsistent naming

❏ Missing, inaccurate, or outdated information within the asset records

❏ Lack of a logical hierarchy relating asset records to each other in the EAM or CMMS

❏ Lack of accurate and complete equipment history

❏ No ability to assess asset life-cycle costs or make good repair vs. replace decisions

❏ No way of definitively knowing whether or not a maintenance procedure is working effectively

❏ Inability to calculate maintenance spend and productivity losses at the asset level

Ensuring asset data integrity, therefore, becomes important not only to the maintenance organization but also to all facets of the plant organization, and even the entire company.

12.3 Assessing Material Data

Another major class of data related to physical assets includes the spare parts or general materials used for repairs to the assets. Many spare parts are typically kept in stock to support part replacement. In addition to the asset-specific spare parts, there is often an extensive inventory of non-asset-specific maintenance repair materials. The items in this latter inventory category are often referred to as generic items. These items include parts like fasteners, pipe and pipe fittings, cable, and conduit. These items are typically items that can be used on multiple kinds of assets. The financial value of spare parts and generic parts inventory is usually very high—often tens of millions of dollars.

The items kept in inventory are used throughout the fiscal year to support asset repairs. The cost of each part withdrawn from inventory is charged to a maintenance expense account as it is used, credited against the value of the inventory as that inventory is depleted. When minimum inventory levels for an item are reached, a purchase order is issued to replenish that inventory. But the expense for the withdrawn part is charged to the maintenance budget. Aside from material items coming from inventory, during a typical year, some items are purchased from outside sources. This can occur when the needed item is not kept in inventory or when the worker needing the part cannot find it in inventory. (The part may be there, but the data used to describe it is un-standardized). Whatever the reason for outside purchases of material items, their expense is also charged to the maintenance budget. The amount of annual maintenance expense related to material items usage is comparable to the cost of maintenance labor. Both are very high in a typical industrial plant—again each potentially in the tens of millions of dollars.

According to empirical data from hundreds of plants, the spare parts portion of total maintenance expense ranges from 40–60% in developed countries. In countries with significantly lower labor costs, the split is even more heavily weighted toward spare parts costs. Beyond costs, the amount of time spent searching for and obtaining needed materials—either in a catalog or physically in a storeroom—is always prolonged when the data representing those items in the inventory catalog is of poor quality and inconsistent. For these and other reasons, concentrating on the quality of materials data is essential.

12.4 Data Strategy Session

The purpose of conducting a thorough asset data or material item data assessment is to establish an understanding of the condition of the existing, or legacy master data. This data will be used to populate a new computer system or simply needs a major clean-up to make it usable within the existing system. In either case the effort to transform or clean up the legacy data is a large undertaking. For this reason careful planning is advisable.

A data strategy session should be conducted to develop an understanding of the business objectives of a proposed data transformation or clean-up effort. This understanding is necessary to create a to-be data integrity model that will support the business objectives.

12.5 To-Be Taxonomy

Before the actual review of the legacy data can begin, an understanding of the to-be format is essential. What do you want the data to look like when you are finished? You need a frame of reference against which to evaluate the condition of the data. Most often legacy data is not in as good a condition as you would wish. Work will be required to get it ready for its future home.

As described in Chapter 11, a taxonomy defines the classification system that will be used to describe assets or material items. There are existing taxonomies for asset and material item data, typically available from specialty consulting firms. It is advisable to contract for such existing taxonomies as these have been created over time and are proven classification systems for assets and material items. If such an existing taxonomy is adopted as an advanced starting point, some tuning may still be required to ensure support of the business objectives which were defined in the data strategy session. Nevertheless, tuning an existing taxonomy is orders-of-magnitude easier than trying to create one from scratch.

Perhaps the best way to describe a to-be taxonomy for assets and material items is to provide examples. The following excerpts are from existing asset and material taxonomies.

Asset Taxonomy

This taxonomy describes an electric AC motor. The elements of the taxonomy consist of the primary bullets in the list below. The primary bullets are the metadata (remember—the metadata is the data about the data, an explanation of what is required in each field). Examples of the kind of data that may actually be entered into each field for a particular asset are provided in italics.

❏ Class, Subclass (The item noun name and the adjective that distinguishes it)
  • Example: *Motor, Electric AC*
❏ Manufacturer
  • *Example: Tesla Induction Motor Company*
❏ Model Number
  • Example: *X7A*
❏ Serial Number
  • Example: *D143218*
❏ Attribute / Specification Template (note that the attributes will vary by class/subclass)
  • HORESEPOWER: *Example: 10*
  • PHASES: *Example: 3*
  • HERTZ: *Example: 60*
  • RPM: *Example: 1150*
  • VOLTAGE: *Example: 460*
  • AMPERAGE: *Example: 12.7*
  • FRAME: *Example: 256T*
  • DUTY: Example: *Continuous*
  • INSULATION CLASS: *Example: F*
  • ENCLOSURE TYPE: *Example: TEFC*
  • SERVICE FACTOR: *Example: 1.0*
❏ Asset Description (typically a stringing together of various attribute metadata and entered attributes)
❏ Example: Motor, Electric AC, 10HP, 3 Phase, 60HZ, 1150RPM, 460V, 12.7Amps, Frame 256T, Continuous Duty, Insulation Class F, TEFC Enclosure, 1.0 SF, Tesla Induction Motor Company Model X7A, S/N D143218

A typical existing asset taxonomy would consist of several hundred entries, with anywhere from 100–300 being used in a typical asset master file, depending on the size and complexity of the facility in question. Note that the taxonomy entry for Motor, Electric AC (the example above) would count as one entry in the taxonomy.

Material Item Taxonomy

The following material taxonomy describes a ball bearing. Recall this same example was used in Chapter 10. The elements of the taxonomy consist of the primary bullets while examples of data that may be entered into each field for a particular material item are provided in italics.

❏ Class, Subclass (The item noun name and the adjective that distinguishes it)
  • Example: *Bearing, Ball*
❏ Manufacturer/Supplier Name
  • Example: *SKF Bearing*
❏ Manufacturer Part Number/Supplier Part Number
  • Example: *60032RS*
❏ Alternate Manufacturer/Supplier Name(s)
  • Example: *Fafnir Bearing*
❏ Alternate Part Number(s)
  • Example: *9103PP*
❏ Attribute / Specification Template (note that the attributes will vary by class/subclass)
  • ROW: Example: *Single*
  • THRUST: Example: *Radial*
  • SERIES: Example: *Extra Light*
  • SEAL: Example: *2 Contact*

- I D.: Example: *17 MM (0.6693 IN)*
- O.D.: Example: *35 MM (1.3780 IN)*
- WIDTH: Example: *10 MM (0.3937 IN)*
- ❑ Material Description (typically a stringing together of various attribute metadata and entered attributes)
  - Example: Bearing, Ball, Single Row, Radial Thrust, Extra Light Series, 2 Contact Seal(s), 17 MM (0.6693 IN) I.D., 35 MM (1.3780 IN) O.D., 10 MM (0.3937 IN) Width, SKF Bearing P/N 60032RS, Fafnir Bearing P/N 9103PP
- ❑ Additional information typically not found in the description:
  - ❑ Commodity Code
  - ❑ Company ID number
  - ❑ Location Code

A typical existing material item's taxonomy would consist of several thousand entries, with anywhere from 1,000–1,500 being used in a typical material items catalog. The entry for bearing, ball (the example above) would count as one entry in the taxonomy.

12.6 Primary Data Fields

As the above to-be taxonomy excerpts show, describing an asset or a material item requires a wide variety of attributes. Some of the attributes are considered more important than others. These are referred to as the primary fields. Primary fields are of greater interest in assessing data integrity for various reasons, including the fact that primary fields usually are needed in order to know what the other attribute fields should be. Your target benchmarks in the various data quality dimensions (DQDs) are usually more stringent for primary fields than for other fields. For example, the DQD completeness (also known as fill rate %) has to be higher for primary fields than for other fields.

When we move beyond primary fields, all other fields are not of equal importance. An order of importance would dictate a higher standard of quality for some fields over other fields. A good taxonomy will prioritize the attributes in order of importance. For example, some attributes are essential in order to readily find an asset in a large asset file.An incomplete description of an asset could result in the wrong asset being selected for a repair request by an equipment operator, and the maintenance department not knowing which asset to go fix. Some attributes are essential in order to readily find an asset in a large asset file, and readily purchase it from a supplier. An incomplete description of a material item could result in the wrong item being purchased or the supplier not knowing which exact item you want.

Other attributes may be nice-to-have as additional information, but not essential for finding an asset or purchasing the material item. We will provide some insight into primary fields, and the prioritization of other fields, in the following sections.

12.7 Class and Subclass

The class is the noun name for an asset or for a material item. An example of an asset *class* is *Pump*. An example of material item *class* is *Screw*.

The subclass represents the adjective that distinguishes an asset or material item from similar assets or material items of the same class. To carry our examples further:

❑ Within the asset class of Pump, the subclasses *Centrifugal* and *Reciprocating* distinguish these two kinds of Pumps from each other, and from other kinds of Pumps.

Class	Subclass
Pump	*Centrifugal*
Pump	*Reciprocating*

❑ Within the material item class of screw, the subclasses *Wood* and *Machine* distinguish these two screws from each other and from other screws.

Class	Subclass
Screw	*Wood*
Screw	*Machine*

Class and subclass are considered primary fields for asset records and for material item records. The assignment of the class and subclass determines which entry in the overall taxonomy should be called (in other words, which metadata applies to this item). As you can see, the metadata for a centrifugal pump is different from the metadata for an electric motor, and similarly the metadata for a ball bearing is far different than that of a machine screw.

To assess the master data for the data fields *class* and *subclass*, apply all the relevant DQDs to each of these data fields. Pay particular attention to the DQDs completeness (fill rate %), consistent representation, free-of-error, interpretability, understandability, and value-added (refer to Chapter 11 for a list of DQDs and their definitions).

The following simple assessment model will give you an idea of what you should be looking for during the assessment of class and subclass in asset or material item master data.

❏ Fill rate (class) = what is the percentage of line items in the legacy data that have class data populated (regardless of whether the class conforms to the to-be model)
  • Calculate the percentage of line items that contain a *class*
  • Note the actual number of line items (filled and not filled)
❏ Fill Rate (subclass) = what is the percentage of line items that have both a class and a subclass populated
❏ Taxonomy Conformance = of the line items that have a class or a class/subclass combination, what is the percentage that conforms to the to-be taxonomy?

It is important to note here that legacy systems may not have included discrete fields for class and subclass. Equivalent fields may be present that could serve the purpose of class and subclass. For example some systems refer to a *noun* and *qualifier*. These may be direct proxies for class and subclass. In cases where there are not direct or proxy fields in the legacy system for class or subclass, it may be possible to derive the class and subclass from an intelligent numbering system present in the legacy data. For example,

❏ The asset number "P101A" may signify that this asset is a pump because the *"P"* part of the number signifies pumps.
❏ The material item number of 063-12345 may signify that this material item is a bearing because the *063* part of the number signifies bearings.
  For both assets and material items, there are cases where the class and/or subclass could be derived from the legacy description (vs. an

intelligent number). The old description may include a noun name that could be extracted and used for the class field in the to-be data model.

In these examples, the fill-rate of legacy data should be calculated based on the *initial* fact of whether there is a class and subclass populated. A low initial fill-rate indicates the need for some of the activities described in the previous paragraph to get to an acceptable fill-rate level.

Just because data is populated in a given field does not mean it is error-free. Misclassification, spelling errors, and formatting errors will be common. Fill rates determine only if the data field is populated. For example, *TANK* and *VESSEL* may appear as asset classes and SCREW, BOLT, and STUD may appear as material classes in the approved taxonomy. Although these terms, technically speaking, should apply to different classes of assets and material items, they are sometimes used interchangeably. Consequently misclassification is very common.

Let us use our asset example above to explore this concept further, understanding that the asset example equally applies for material items.

Tank and vessel are often used interchangeably, but technically they are different kinds of assets—both are containers, could have similar characteristics, and sometimes have similar functions. However, technically, vessels are containers under pressure, whereas tanks are not (tanks may be covered or open to the atmosphere but they are not containers under pressure). Significantly different operating and maintenance procedures apply to each. The technically correct classification may not be obvious to individuals responsible for classifying this asset if they are not subject matter experts. Subject matter expertise is needed to make these distinctions—thus, the importance of involving resources beyond the IT resources.

If proper classification and sub-classification are in question, they can sometimes be verified using the asset's manufacturer name, model number, and serial number, or the material item's manufacturer name and part number. These numbers can usually be used to view the specifications of the asset or material item in the manufacturer's technical documents. These documents are often available on line or as published documents. These sources will usually provide insight into what the class and subclass should be.

Inconsistent representation means there are variations from standard naming conventions in the class or subclass field. For example,

❑ PMP verses PUMP

❑ BRG verses BEARING

Abbreviations are one common cause of inconsistent representation. Legacy systems often have limited field lengths requiring abbreviations, but abbreviations can vary from asset to asset, and from material item to material item. You should measure the approximate percentage of legacy line items that contain these variations. This measure will indicate work that needs to be done to standardize the entries. Regional geography and industrial terminology can also be a factor with inconsistent representation, especially when consolidating disparate systems into a common technology platform.

To enhance the DQDs of *interpretability* and *understandability,* the class and subclass should always be identified by the asset's or material item's physical characteristics, and not by the use or application of the asset or material item. For example, *reciprocating* and *centrifugal* are acceptable subclasses of the asset class pump because they describe the physical characteristics of the *pump*. However, *water* or *chemical* should be avoided as subclasses for *pumps* because these relate to the use or application that may vary depending on where this asset is being installed.

Because class and subclass are primary fields, a benchmark for their final fill-rate is 100%, meaning that all of the assets in the master asset file, and all of the material items in the master materials catalog, must eventually have a proper class and subclass assignment. If the initial fill-rate is below 100% activities will be necessary to bring the fill-rate up to benchmark standards. Remember, the assignment of the class and subclass defines all of the rest of the attributes needed to properly classify and describe an asset or a material item.

During an assessment, having the to-be taxonomy and data model is essential to being able to evaluate the readiness of the legacy data to fit that to-be taxonomy. Sometimes data needed for discrete fields in the to-be data model are readily available in their own discrete fields in the legacy system. When this is the case the legacy fields can be mapped directly to their related fields in the new system.

However, much of the needed attribute information is buried within larger description fields in the legacy records. These description fields (we'll refer to them as *old descriptions*) are often large text fields in the legacy systems that contain mostly unformatted attribute information. When that is the case, the data needed for discrete fields in the new system has to be extracted from these old descriptions. Extraction of discrete pieces of data from old description fields is a cumbersome task if done manually, but can be eased with the use of specialty tools designed for this purpose. In either case, the needed attribute data must be located in the old description field and entered into a discrete field of its own. Then these discrete fields can be used to populate attribute specification templates in the to-be model.

12.8 Manufacturer or Supplier Name

For assets and material items, the manufacturer name is simply the name of the company that originally manufactured the asset or the material item (also known as the Original Equipment Manufacturer or OEM). In all cases, manufacturer name is considered a primary field. However, material items have some additional variables that must be taken into account. For material items, the manufacturer name could be the name of the company that purchased the part from the OEM for a larger assembly, or the name of the company that acts as a retail or wholesale distributor for the part. In this latter case, the company is referred to as the *supplier*, and they may assign their own numbers to their items.

Apply all the relevant DQDs to the manufacturer name field. Again, for manufacturer name, pay particular attention to completeness (fill rate %), consistent representation, free-of-error, interpretability, understandability, and value-added. Inconsistent spelling, formatting, and missing data may be prevalent throughout the legacy data. Best practices for manufacturer names include:

❑ Use the company's full legal name, and establish a consistent spelling and format so that there is only one representation for each company.

❑ Ensure there is at least one manufacturer name for every asset in the master asset file.

❑ Ensure there is at least one manufacturer name and/or supplier name for every material item in the catalog. There are instances where the manufacturer name will not be applicable. For example, a manufacturer name may not be required for fasteners (nuts, bolts, screws) because these parts are typically purchased using other attribute data. However, when the manufacturer name is not appropriate, at least one supplier name should be assigned for material items.

The following charts illustrate the importance of standardizing manufacturers' names.

Assets

Un-standardized Manufacturer	Standardized Manufacturer
GE	General Electric Company
General Electric	General Electric Company
Gen Elec	General Electric Company
G.E.	General Electric Company

Material Items

Un-standardized Manufacturer	Standardized Manufacturer
A Bradley	Allen Bradley Inc
Allen Bradley	Allen Bradley Inc
AB	Allen Bradley Inc
Bradley Inc	Allen Bradley Inc

As may be probably obvious from the charts, a computer system would consider each of the eight un-standardized entries to be unique, when in reality these eight entries represent only two real companies. A search for a General Electric Company asset that specified the search criteria to be "GE" would not retrieve any of the assets that had the other three spellings assigned.

A target benchmark for the attribute manufacturer name fill rate is 90% for assets, meaning that almost all of the assets in the master asset file must eventually have a proper *manufacturer name* assignment. For material items, the target benchmark for fill rate is also 90%, however some material items will have a supplier's name in place of a manufacturer's name.

Author's Note

When we get past Class, subclass, and manufacturer name, the differences between assets and material items have to be addressed. Section 12.9 relates to data that is specific to assets whereas Section 12.10 relates to data that is specific to material items. There are similarities between the two types of data, yet they are different and often handled by separate departments within a company or plant. Because of the similarities, the reader may notice repetition of some of the wording between Sections 12.9 and 12.10. The reason for this repetition is to provide the same level of detail for those readers interested only in asset data as for those interested only in material items data.

12.9 Asset—Model Number or Serial Number

The model number and/or serial number are also critical components of a complete asset description because they uniquely identify a given asset from a manufacturer. Model number and serial number are considered primary fields for assets.

All CMMSs and EAMs contain specific fields for a model number and a serial number in the asset record. The assignment of these numbers in the legacy system was probably done during the initial setup of the asset record without the benefit of a controlling rule set.

Apply all the relevant DQDs to each of these data fields. For model number and serial number, you should again pay particular attention to completeness (fill rate %), consistent representation, free-of-error, interpretability, understandability, and value-added.

It is best practice to have both the model number and serial number for each asset, assuming they are available. Having both is essential to understanding the inner workings of the asset, to obtaining the correct spare parts when needed, to understanding the proper maintenance procedures, and to reordering the asset when it reaches the end of its useful life.

One area of concern is model number or serial number duplication. Manufacturers use essentially two styles of model numbers and serial numbers:

❏ Indicative—format-driven identification, intelligent (the character string has meaning)
❏ Non-indicative—could be sequential, non-intelligent (the character string has no meaning)

The model number is intended to be a unique identifier for a *type* of asset whereas the serial number is intended to be a unique identifier for that *specific* asset. You can typically have many assets in an asset master file with the same model number, but you should not have more than one asset with the same serial number. The more complex the serial number character string, including the length and the combination of alpha and numeric characters, the more unique it will be compared with the other numbers in the master data. Shorter serial numbers are more likely to have some duplication, meaning a different manufacturer may have coincidentally assigned the same number to a different asset.

As mentioned above, if two or more assets in the asset master file have the same model number and the same manufacturer name, they are probably valid duplications. It is common for a particular model pump to be used in many locations throughout a facility. However, if two or more assets have the same model number, but *different* manufacturer names, that would probably indicate that it is a different kind of valid duplication—in this case, two different manufacturers coincidentally used the same model number for different assets.

For serial number, valid duplication is less likely but still possible. Generally speaking, there should not be many valid duplicate serial numbers in an asset master file. The only ones that would be valid are when two manufacturers coincidentally used the same serial number for two different kinds of assets. These would be *valid* duplicates. As a measure to deal with valid duplicate serial numbers, some to-be data models include a created field that combines the manufacturer name with the serial number, effectively creating a unique asset identification number. There should not be any duplications of the combination of the manufacturer and serial number in an asset master file.

Here are some examples of additional common model and serial number errors that may be uncovered with a good data assessment:

❑ Inconsistent formatting: X7A versus X-7A
❑ Incomplete information: X7 versus X7A
❑ Using upper case for alpha characters inconsistently
❑ Missing model or serial numbers
❑ Model or Serial numbers not in the correct data column
❑ Having a number misrepresented as a model number or serial number
❑ Obsolete numbers
❑ Duplicate numbers that have not been validated

Assessing the model number and serial number master data requires applying all the relevant DQDs to each of these data fields. For model number and serial number, pay particular attention to the DQDs completeness (fill rate %), consistent representation, and potential errors.

A target benchmark for the attributes model number and serial number fill rate is 90%, meaning that almost all of the assets in the master asset file must eventually have a proper model number and serial number assignment.

12.10   Material Items—Manufacturer or Supplier Part Number

The part number is also a critical component of a complete material item description because it uniquely identifies a given part from a manufacturer or supplier. Manufacturer or supplier part numbers are considered primary fields.

All CMMSs and EAMs contain at least one (usually several) specific field for a manufacturer or supplier part number. The assignment of a manufacturer's or supplier's part number in the legacy catalog was probably done during the initial setup of the item record without the benefit of a controlling rule set. Therefore, it could be incomplete or missing altogether. Another common problem encountered when setting up a part in a computer system is determining which part number should be entered into the computer system—the manufacturer's part number, the supplier's, or both? This is a particular problem when the software application does not have the ability to accept multiple part numbers. In many cases, even when the system does allow multiple part numbers, the initial set-up effort may have taken short-cuts or

done simple data conversions, resulting in the wrong numbers residing in the wrong fields.

Apply all the relevant DQDs to each of these data fields. For manufacturer or supplier part numbers, you should again pay particular
attention to completeness (fill rate %), consistent representation, free-
of-error, interpretability, understandability, and value-added.

It is best practice to have at least one manufacturer part number
or one supplier part number for each material item, assuming they are
available. If both are available, they should both be populated in the
material item master record. There are many instances where there
will not be a manufacturer part number. In these cases, the supplier
part number is essential to the replenishment and inventory control
processes.

One area of concern is manufacturer or supplier part number
duplication. Like manufacturers for assets, manufacturers and suppliers of material items use essentially two styles of part numbers:

❏ Indicative—format-driven identification, intelligent (the character
   string has meaning)
❏ Non-indicative—could be sequential, non-intelligent (the character
   string has no meaning)

The manufacturer or supplier part number is intended to be a
unique identifier for a material item. The more complex the part number character string, including the length and the combination of
alpha and numeric characters, the more unique it will be compared
with the other part numbers in the master data. Shorter part numbers
are more likely to have some duplication, meaning a different manufacturer may have coincidentally assigned the same number to a different item.

If two or more items have the same manufacturer or supplier part number, but are from different manufacturers or suppliers, and have different technical attributes, they are most likely valid duplications. Therefore, we need to have both items represented in the catalog because they are different items, even though they have the same part number. There are bound to be duplications simply because there are a finite number of alpha-numeric part number combinations that could be used across all industries and manufacturers, especially when you consider that many part numbers have fewer than eight characters.

As a measure to deal with *valid* duplicate part numbers, some to-be data models include a created field that combines the manufacturer name with the manufacturer part number, effectively creating a unique catalog identification number. These unique catalog identification numbers are sometimes referred to as the CATID.

Some duplications are not different parts and, therefore, are not valid duplicates. Instead *invalid* duplicate part numbers are redundant representations in two or more records of the same item. Invalid duplications exist in legacy master material items data for many reasons, including:

❏ Two or more disparate computer systems have merged

❏ Two or more plants or business units have merged their data

❏ Two or more companies have merged through acquisition

❏ No standard taxonomy or master data rules are in place to control the set-up of new items in the materials catalog

❏ Personnel responsible for the creation of material items master data have not been sufficiently trained in the taxonomy or rule-set

❏ Quality control procedures are lacking or do not exist

These reasons account for most of the invalid duplications. Unlike the valid duplications, invalid duplicate part numbers need to be identified, evaluated, and consolidated in the to-be data model.

Here are some examples of additional common part number errors that may be uncovered with a good data assessment:

❏ Inconsistent formatting: 6203-2Z-NR versus 6203-2ZNR
❏ Incomplete information: 6203 versus 6203-2ZNR
❏ Using upper case for alpha characters inconsistently
❏ Part numbers associated with incorrect suppliers or manufacturers
❏ Missing part numbers
❏ Part numbers not in the correct data column
❏ Having a number misrepresented as the part number: serial number, model number, etc.
❏ Obsolete part numbers
❏ Duplicate part numbers that have not been validated

We talked earlier about a CATID, which is in essence an internal company-assigned number created by combining various data elements together to create a unique identifier. In our earlier example of valid duplications, two different manufacturers unknowingly assigned the identical part number to two different items. By combining each manufacturer's name (or a code for their name) with the duplicate part number, you can render two new unique identification numbers. Sometimes more than just the manufacturer field and the part number field are combined to create a unique CATID.

Like manufacturer part numbers, two types of CATID part numbers are used:

❏ Indicative—format-driven identification, intelligent
❏ Non-indicative—sequential, non-intelligent

An example of how an indicative CATID could be created follows:

Using a car oil filter at the local parts store as an example, the first field could include the manufacturer name, the second field could include the filter media, the third field could include diameter, and the fourth field could include a simple sequential number. Refer to Figure 12-1 for an example of an indicative CATID using this example.

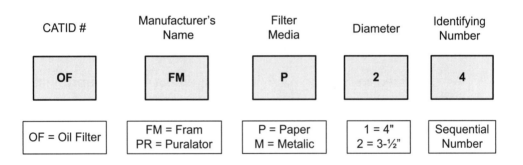

CATID #	Manufacturer's Name	Filter Media	Diameter	Identifying Number
OF	FM	P	2	4
OF = Oil Filter	FM = Fram PR = Puralator	P = Paper M = Metalic	1 = 4" 2 = 3-½"	Sequential Number

Figure 12-1 An Example of an Indicative CATID

The shaded boxes in Figure 12-1 contain the newly-created CATID.

Examples of a non-indicative CATID could be 10000, 10001, 10002.

Assessing the part number master data requires applying all the relevant DQDs to each of these data fields. For manufacturer or supplier part numbers, pay particular attention to the DQDs completeness (fill rate %), consistent representation, and potential errors. In addition, if CATIDs are found or created, compare them to the manufacturer and supplier part number for consistent use and format. Exceptions should be noted for resolution.

A benchmark for the attributes manufacturer and/or supplier part number fill rate is 90%, meaning that almost all of the material items in the catalog must eventually have a proper manufacturer or supplier part number assignment, or both if available. As mentioned earlier, many items do not have manufacturer part numbers (e.g., pipe fittings), but in those cases you should strive to have a supplier part number if possible.

12.11   Attribute Templates

As discussed in Section 12.7, the assignment of the appropriate class and subclass entry from the to-be taxonomy defines the variable attribute fields that apply to that class and subclass of an asset or material item. Remember that the attributes for a centrifugal pump are different from the attributes for an electric motor and a ball bearing has different attributes from a machine screw.

In many legacy asset master data files or material item master data files, there are no discrete fields for the variable attributes. But the data may still be present in other fields, including legacy description fields. In those cases, the assessment needs to determine the extent to which the data for these fields exist. If individual fields exist in the legacy data, then it may be easier to map and convert that data into the to-be data model. If some or most of the needed data is buried in other fields, then more work will be needed to extract that data, format it, and then populate it into the to-be data model.

Attribute template fields are not generally considered primary fields. However, some of the attributes may be absolutely necessary to be able to adequately describe an asset or a material item. Each class/subclass entry in the taxonomy will have associated with it the appropriate attributes that are necessary. An asset and material item example is appropriate at this point.

Asset

The following is the electric motor attribute template we used earlier in this chapter. The primary fields are designated along with the relative importance of the other fields in the attribute template also designated:

❏ Class, Subclass: Motor, Electric AC—*Primary Field*

❏ Manufacturer Name—*Primary Field*

❏ Model Number—*Primary Field*

❏ Serial Number—*Primary Field*

❏ Attribute/Specification Template for electric AC motor—*Not Primary*

- HORESEPOWER—*Important*
- PHASES
- HERTZ
- RPM—*Important*
- VOLTAGE—*Important*
- AMPERAGE
- FRAME—*Important*
- DUTY
- INSULATION CLASS
- ENCLOSURE TYPE
- SERVICE FACTOR

Material Item

The following is a ball bearing attribute template. The primary fields are designated along with the relative importance of the other fields in the attribute template.

❏ Class, Subclass: Bearing, Ball—*Primary Field*

❏ Manufacturer/Supplier Name—*Primary Field*

❏ Manufacturer Part Number/Supplier Part Number—*Primary Field*

❏ Alternate Manufacturer/Supplier Name(s)—*Not Primary*

❏ Alternate Part Number(s)—*Not Primary*

❏ Attribute / Specification Template for ball bearing—*Not Primary*

- ROW
- THRUST—*Important for finding and purchasing*
- SERIES
- SEAL
- Inner Diameter—*Important for finding and purchasing*
- Outer Diameter—*Important for finding and purchasing*
- WIDTH

❏ Material Description (typically a stringing together of various attribute metadata and entered attributes)

Keeping in mind that there are two to three hundred entries in a typical asset taxonomy (*again, motor, electric AC is one entry*), and two-to-three thousand entries in a typical material item taxonomy (*bearing, ball is one entry*), you can appreciate that the designation of the relative importance of the various fields contained in each entry would be a lengthy and time-consuming process. Happily, the existing taxonomies that have been created by the better specialty consulting firms already have assigned the relative importance designations for the thousands of entries in their taxonomies. As you can imagine, these designations are extremely helpful in assessing legacy data. They provide the reference point against which to conduct the assessment. If legacy data is missing or lacking for primary fields, it is obviously a more important finding of the assessment than missing data for less important fields.

Assessing the readiness of the legacy data to ultimately achieve good fill-rates for the primary and important fields requires manual viewing of the legacy data if a specialty software application is not available. Many assessments are conducted using rudimentary tools like spreadsheets. In this kind of scenario, the manual viewing effort includes viewing the legacy description fields and viewing discrete fields in a spreadsheet-like view. Subject matter experts are usually needed to do this effectively. But even with dedicated subject matter experts, the assessment process is very manual and could be so time-consuming as to be prohibitive.

If the last paragraph depressed you, take heart! The specialty consulting firms have developed software tools which, when coupled with their pre-existing taxonomies and subject matter experts, can automate much of this review process. The software tools of the better firms have tens of thousands of pre-developed rules that can be used to conduct automated review of various fields in the legacy data. These tools include ones that are designed to look for recognizable strings of characters that are likely to be data elements needed in the to-be data model. An example is a software application that is designed to scan large volumes of legacy data searching for any strings of characters that might match the following list: GE, G.E., General Electric, General Electric Company. Anytime the software finds such a string, it suggests to the assessor that these are likely manufacturer names. The software can further be programmed to actually populate the to-be manufacturer name field with a standard spelling of the General Electric Company. These tools can make short work of assessing the legacy data, and to some extent actually enhancing the legacy data to advance it to a more ready state for the to-be data model. Strong consideration should be given to engaging a qualified specialty consulting firm to assist in the assessment and enhancement process.

12.12 Other Asset Data Fields

Other specific asset-related fields that should be analyzed for fill rates include Parent ID, Physical Locations, applicable Cost Center codes, and a hierarchical Functional Location reference.

Parent ID fields are used to relate an asset to another asset. A good example of this is relating a motor with a pump. Because the motor is connected to the pump and is there specifically to drive the pump, then it makes sense to want to see these two assets related to each other in the database.

Physical Location fields typically are used to provide information that helps locate the asset in a large facility. Sometimes physical location references use a Building/Floor/Column schema (e.g., Building 5, Floor 3, Column A-4).

Cost Centers are normally codes provided from the accounting department indicating an area of a facility where the asset resides. For example, the cost accounting code structure may be designed to track costs in different areas of a facility. Assets that reside in a particular cost center can be designated as such by entering that cost center code in the appropriate field. Once this is done any work orders that are subsequently opened against a particular asset will automatically roll-up the costs of the work order to the correct accounting cost center.

The *Hierarchical Parent Location* field is usually a reference to a functional location record in a larger table of functional locations records. Functional location records are not to be confused with other asset records. The functional location hierarchy serves to group assets together into systems and/or areas of a facility, allowing the assets that reside at these functional locations to be organized into logical groupings.

To visualize what is meant by a functional location hierarchy, think of a family tree or an organization chart. Using the organization chart example, each department manager in a company is depicted as an individual box on that chart, with lines connecting each box up to the manager's box above, and so on. Eventually all boxes "roll-up" to the CEO box at the top of the organization chart. In the same way, a plant or facility can be organized in such fashion. Each area of the plant is represented by a functional location record (a box on the chart). Each record (box) is connected to the next higher level by entering the parent functional location number. Eventually all functional locations roll up to the top functional location which may represent the entire plant.

Asset records are not functional location records. However, in most EAM or CMMS, asset records can be assigned to a functional location record, placing that asset into the part of the functional location hierarchy where it belongs. Think of assets as employees in an organization chart that may rove from department to department over time. When this happens, the previous department remains in place and unchanged, except that the employee no longer reports into that department.

Like employees changing departments, assets can be physically moved into and out of different functional locations over time. When this happens, the moves must be reflected in the EAM or CMMS data. This is done by simply changing the functional location reference field in the asset record. We separate functional locations from asset records because of the fact that the functional configuration of a plant does not change frequently (unless capital modifications or plant expansions are done) whereas assets do tend to move into and out of functional locations. This separation eases the data maintenance burden—when the asset moves away, the functional location record does not need to change. The only necessity is to change the functional location reference in the asset record.

Functional location reference fields in asset records are considered primary fields and should have a fill rate percentage above 95%.

12.13   The Goal—Quality Data for the Future

When evaluating data—whether formally through an assessment or informally through a user review—there are going to be data inconsistencies, lack of data, or misplaced data. There are also various levels of sophistication and technique for conducting data assessments. A simple assessment may include checking for null data in a given field of a record. A more sophisticated assessment may involve evaluating the data against specific DQD models. The more detailed the assessment, the more sophisticated the analysis must become—and, in turn, the longer the assessment will take, particularly without the right software tools to do the analysis. The goal of the assessment is to identify and report data irregularities and gaps in business processes so that they can be corrected. You want to ensure that the data assessment results convey meaningful actions that should be taken from a business case perspective.

Chapter 13

Asset Data Clean-Up and Repair

Problems that are not remedied only get worse.

Unknown

13.1 After the Assessment

Data integrity assessments—whether they focus on assets or material items related to the assets—are valuable exercises. They show you the gaps and flaws in your data integrity and help you recognize weaknesses that could easily lead to flawed business decisions. However, undertaking a data integrity assessment is an absolute waste of time unless you respond actively to what you discover as a result of the assessment. The problems uncovered can span a wide variety of issues. In this chapter, we will focus our attention on data clean-up and filling the data gaps that typically are uncovered from an asset data assessment process.

13.2 Data Repair is Far from Simple

Data repair is not a simple task. Before we begin that discussion, however, let's clearly understand why assessment and repair are both required. Whatever approach you take to clean up data problems, the activity associated with the clean-up will be highly visible to the organization. The reason for this is that a clean-up effort is something very different than the normal day-to-day activities within an organization. Equally visible to the people in an organization is the decision to do nothing once an assessment has been completed and gaps have been identified.

Every activity has an impact on the organization. What message do you send when you do an assessment but, for whatever reason, do not make the data repairs? The message is clear: Asset-related data has little value. With that message, the organization will accept the poor data quality, continue to compensate for the lack of data integrity, and consider it to be of little value to try to improve or maintain data quality. On the other hand, if you do act on the knowledge about data quality obtained through a data assessment, the organization will react differently. If you follow through with the corrections, you will reinforce that data integrity is valuable and organizational support is more likely to keep correct into the future what you have fixed.

13.3 Repair Problems

The vast majority of organizations do not have only one single integrated asset or inventory database. Instead they have many disconnected stand-alone databases, each owned by a different functional group. You need to be aware that you may not be fixing a single database, but rather all of the databases that contain asset or materials information. Failure to recognize this—especially when there is no intention to replace the myriad legacy systems with one fully-integrated, enterprise-wide system—can easily result in wasted effort and rework, both of which will cost time and money. Many gaps may typically be uncovered during a data assessment.

Missing Data

It is important that asset records have its data elements populated and that the data are formatted consistently across all of the affected databases. The fact that these databases are often owned by different groups can make this difficult. However, it is nonetheless necessary to address all asset and materials data wherever it resides. Not every data element in an asset or materials database must be populated with data to support business functions. The fields that are important to fulfill business goals do need to be populated with high-quality data. The decision as to which fields are critical and which are not should be left to functional and subject matter experts.

Inaccurate Data

The problem with inaccurate data is that you don't know it is inaccurate until you try to use it and discover it is in error. An assessment can determine the level of accuracy by allowing you to analyze a statistical subset of the data. This analysis will help you estimate the overall degree of accuracy of the full set of data. The estimated total overall accuracy should help you determine the clean-up strategy. A low level of accuracy may dictate the need for a project in which all asset data are reviewed and corrected, whereas a high level of accuracy may dictate a less extensive intervention.

Inconsistent Data Across Multiple Databases

Many companies have asset and materials data in multiple databases. Some companies have integrated systems where the master asset and materials data reside, but that data are used by other systems requiring such asset and materials data. In either case, the multiple applications originated at different times and were driven by different functions within the organization (Maintenance, Operations, Engineering, Inspection, Capital Construction and others). Although they may all have the same data about the same assets and the same material items, the assets and material items are often described differently and entered into the systems in different formats. Examples of these differences include:

- ❏   Different descriptions of key data elements
- ❏   Punctuation, dashes, capitalization, abbreviation, and other spelling differences
- ❏   Disagreeing data about the same asset or material item across multiple applications

These problems and others are further complicated depending on the number of applications involved and the fact that you may not know which one holds the correct data. An even larger problem is encountered when the need to align and synchronize the data across multiple applications meets with the reluctance of the affected departments to change.

Aside from the technical data gaps, impediments can be encountered when endeavoring to achieve high-quality data. Some examples follow.

COST

Data integrity does not come for free. Whatever approach you take to data repair, there is always an associated cost. Every company should recognize the value of accurate data and accept the reality that the cost of doing nothing is so much more expensive.

RESOURCE ISSUES

Similar to cost, whichever approach is chosen to do the required repair, resources will be required. Different approaches can have different kinds of resource requirements. If internal resources cannot be dedicated to data clean-up efforts on a full-time basis, it is usually advisable to engage third-party specialty consultants who can accelerate the clean-up process dramatically. This approach places fewer demands on company personnel, but higher demands on company financial resources. In either case, the internal resource requirements will include some personnel to direct the overall effort and provide needed subject matter expertise. Without a long-term strategic understanding of the value of high-quality data, the organization will find making the needed resources available difficult.

EASE OF REPAIR

Data repair is not easily done in the systems that primarily use the data. These systems tend not to have master data management tools. In a few cases, some rudimentary data management tools are embedded in the applications, but they are not robust enough to do massive data editing. For example, these tools do not have functionality to easily make global changes, to compare data elements to standard libraries, or to track progress in multiple stages of clean-up and status reporting. In examples like those just cited, it also becomes quickly apparent that the data management activities utilizing a spreadsheet are also sorely lacking in the needed functionality.

As we have mentioned throughout this book, there are specialized solutions, tools, and consultants that are specifically designed to address the requirements of data cleansing.

OFF THE SHELF DATA SCRUBBING APPLICATIONS

Companies that have provided data scrubbing services have robust applications with substantial master data management functionality. Very few of these companies focus on physical asset-related data, but there are some which do and have substantial asset and material item content and taxonomies. Some of these data scrubbing tools are not interfaced or integrated to the EAM applications which ultimately use the data. This can cause problems and extra work as you clean up your data. Others do have the ability to extract master data from legacy systems and load it into the system that will ultimately use the data. The applications which have such connectors are also the ones that have physical asset and material item content and taxonomies. These applications would be more desirable choices if a third party is to be engaged for a data assessment and cleansing effort.

DATA CLEANSING SERVICES

Some of the companies which provide the data cleansing software applications also provide professional services for cleansing data, using their tools and content. The expertise of the personnel in these companies varies. Some employ subject matter experts who specialize in asset and materials data. These experts can be relied on to perform the technical work of recognizing the legacy data and knowing how to transform it for the to-be model. Some of the companies employ a model of fully-automated data cleansing, meaning that data is cleansed through the use of low-cost totally automated processes to standardize descriptions. These companies are typically located off-shore and employ non-technical people who do not understand asset and materials data. The use of such companies was attractive for some time because of the significantly lower cost. However, this approach is falling out of favor because the quality of the finished product usually leaves much to be desired. Subject matter expertise is essential in repairing asset-related data. Although a system can support this effort, it is not the ultimate solution.

13.3 Data Repair Strategies

We have alluded to the fact that there are different methods associated with data repair. There are two predominant approaches, one long-term and the other short-term. They approach the work of data repair differently—certainly they take different amounts of time, provide different results over the duration of the effort, and have different cost and resource requirements. They both serve the overall goal of data repair; each has pros and cons. The short-term approach, which is often referred to as the "Big Bang" approach, usually accompanies a dedicated project to clean legacy data all at once. The second predominant approach, the long-term approach, is often referred to as "Fix It As You Go."

13.5 The Big Bang Approach

Description

The Big Bang approach is a shorter-duration project in which all of the asset or material data within a system are reviewed and corrected wherever errors, omissions, or inconsistencies are found. This approach cannot simply be limited to the CMMS database. It needs to address all asset databases including paper files so that at its conclusion all of the data is correct and synchronized. This process will leave you with a clean set of databases in which the organization can have a high degree of confidence.

This approach most often is taken when an organization is moving from one CMMS application to another. In these cases, after spending a great deal of time and money on the CMMS replacement effort, it is preferred to have accurate data from the start. The Big Bang can also be applied to data repair when the existing CMMS will continue to be used. This can be a difficult process because the databases you are working with during the cleansing process are dynamic in nature and have the potential for constant change. However, methods of monitoring changes to data during the cleansing process can be effectively applied to address this difficulty.

Time

Although this approach is referred to as the short-term approach, don't be deceived. Depending on the size of your asset or material database and its condition, extensive activities may be required to bring the data into an acceptable level of quality. Sometimes missing data must be obtained by physically visiting the assets. Sometimes acquisition of the missing data from drawings and manuals may also be necessary. These activities can take time, often many months. Therefore, management must have a long-term strategic view. The effort to attain data integrity needs ongoing support throughout its overall life if the effort is to be successful.

An example of the lack of the long-term view occurred in a project converting from one CMMS database to another. The data cleanup took an extended amount of time, using both internal and external resources. Management did not sustain the strategic commitment and, after several years, lost interest in the effort. At the time management curtailed resources for the effort, the project team was engaged in gathering instrumentation data. The organization decided it would gather the instrumentation data over an extended period as the instruments were repaired. Although the intentions were good, the gathering of instrumentation data never happened. Ten years later, the instrument data was still missing from the CMMS master data tables.

Resources

The Big Bang approach is labor intensive, usually including contractors and consultants. This approach is typically run like any project and requires resources that are skilled in data clean-up. As we described in previous chapters, there are specialty consulting companies that specialize in data clean-up efforts; they bring tools, taxonomy and methods that facilitate their efforts. Even if such companies are engaged, the owner still needs to provide resources to manage the project as well as support the third party team. Internal resources assigned to work with a third-party specialty firm can help find needed information quickly and answer subject matter questions.

Cost

Hiring a specialty consulting firm to assist or lead a data clean-up team is not inexpensive. However, hundreds of such efforts have proven that the efficiencies and dedication of resources attendant with engaging a specialty firm saves time and money in the long run, and usually results in a higher quality result. Although management may believe that internal resources can do the job, studies have shown professionals can do the work three-to-five times faster.

Sustainability

A process needs to be in place to sustain the accuracy and quality of the data once the Big Bang project ends. If this is not done, the best and most accurate data will degrade in quality starting on "day one" following the project's conclusion— changes, additions, and deletions of assets and material items do occur regularly. It is also imperative that these changes, additions, and deletions are captured so that during the months of the big-bang clean-up effort, they will be incorporated and addressed in the master data. One way to sustain the integrity of the data is to require data integrity activities and involvement for every work order or project where assets or material items are altered. Section 14.7 has more on this subject.

13.6 Fix It As You Go

Description

This approach is appropriate if the asset and material items data is already mostly reliable and complete. The "Fix It As You Go" approach is sometimes used only to clean up what is in error. The question is how do you find and then correct asset or materials data that is wrong or missing? One way to find incomplete or missing data is to review that data whenever work is performed on that asset or the asset is replaced. For example, every work order can include a task first, to compare the data recorded in the system with the physical asset itself, next to correct erroneous data, and then to fill in missing data.

It is important to remember that information about an asset or material item may reside in more than one system. As a result, you have an opportunity to ensure that all sources of information for the asset or material item in question are correct. For example, if you correct the information in the CMMS system only, other systems housing data about the asset or material item may still be incorrect, leading to confusion and lots of wasted effort.

Timing

Because you are cleaning and correcting as you go, this approach really has no end point. Some assets or material items may not require repair for many years. Therefore, the data about these assets or material items doesn't need to change. Consider stationary equipment such as tanks, pressure vessels, heat exchangers, and safety valves. In many cases, periodic inspection is mandated by law for such equipment, and occurs over extended periods of time. Other assets may also be on periodic inspection or repair cycles, providing the opportunity to regularly check the integrity of the data during the course of other scheduled activities.

Some asset types may not have regularly-scheduled tasks performed on them. Different regimens are required periodically to verify that the data about these assets remain accurate. (These types of assets are actually less prone to changes, so the risk of data becoming inaccurate is lower.) In any case, you are reviewing data on a piecemeal basis as assets are inspected or repaired. Thus, the timing for the Fix It As You Go approach is indefinite.

Resources

Providing resources for this approach is very different than that of the Big Bang. In this approach many different functions within the organization are involved and play key roles in the success of the overall effort to maintain data integrity. Because every asset's data will be validated when the equipment is inspected or repaired, internal resources across multiple organizations need to understand the ongoing role they play in this process. Some roles within the organization and their possible roles in data integrity follow:

MAINTENANCE PLANNERS

The planners plan and schedule the work that is to be performed on physical assets. They can print out the current data sheets associated with those assets and attach them to the work order so that the repair crew who will be performing the work can validate the information while they are visiting the asset.

MAINTENANCE FOREMEN

When the foremen pick up work orders they need to make sure that the data sheets are attached. These sheets are then given to the work crews to validate the nameplate information while they are in the field.

WORK CREWS

The maintenance work crews are the ones that actually validate the information. Typically, unless given a drawing or a technical manual for use during the repair effort, they may not have access to all of the information on the data sheet they were given. They do have access to whatever information resides on the nameplate of the physical asset and they can at least gather that data. Other information usually resides in documents and drawings that may be filed in different departments. Normally the policy of checking the data as repairs are made does not include allowing time for the repair crews to visit these files and obtain the additional information. For this reason, it is usually the case that such information never gets entered into the asset record. This reality is a disadvantage of the Fix It As You Go approach.

CAPITAL PROJECT ENGINEERS

Those involved with capital projects are the ones that typically handle the installation of new assets or the modification of existing assets. The assets that fall into this category are new assets or assets that are replacing existing assets but have associated with them significant differences—they are not direct replacements-in-kind (replacements-in-kind are typically handled by maintenance crews). It is important that capital project engineers provide to the organization handling data entry the information about the new assets. The proper data can then be entered into the appropriate system.

INSPECTION CREWS

In many cases, a separate department from the maintenance department (often the Inspection Department) has inspectors who are responsible for ensuring the integrity of the stationary equipment. There are legal requirements related to their inspection records. Many of these records include critical data elements associated with assets. The data that the inspectors store often resides in a different data base than the one used by the maintenance department. It is, therefore, the responsibility of the inspection department to make certain that the data for which they are responsible is validated and synchronized with the data in the main CMMS database, as well as other systems which may house information about these assets.

MAINTENANCE ENGINEERS AND TECHNICIANS

Assets are typically the subject of three different levels of maintenance. These are corrective maintenance, preventive maintenance, and predictive maintenance. In many cases, maintenance engineers and technicians are involved in various phases of the work. They are also the ones who work with the equipment files and technical manuals which hold relevant data about the assets. It can be incorporated into their responsibilities to ensure that errors and omissions in the system master data is kept current—specifically that data which resides in the files and data books.

Another asset-related activity involves assets in the rotating equipment asset class (pumps, compressors, fans). Often these assets are disconnected and removed from their operating locations and taken to a shop for repair. These shops typically have drawings and paper files that contain relevant data about these assets. The engineers and technicians in these shops should be responsible for ensuring that the data contained on these drawings and files is provided, as needed, to the CMMS system and maintained as correct and up to date.

Finally, the predictive maintenance technicians regularly visit assets to measure equipment condition. These visits provide another opportunity to validate data. In addition, these technicians maintain technical data about the assets that may be relevant for the CMMS master data. It should be their responsibility to provide accurate and up-to-date information to the people responsible for the master asset data.

OTHER DEPARTMENTS

Many other departments in the plant such as Environmental, Safety, Process Engineering, and Operations use asset-related data every day. They often have databases of their own that could include critical asset information. Because we need to address every asset database as equipment comes out of service for repair, these organizations should not be overlooked. They need to understand their respective responsibilities in the data clean-up effort.

CLERICAL SUPPORT

With all of the data being reviewed, and errors and omissions being identified, there may be a need for a clerical function to enter data into various databases. Technically-trained clerical staff who understand various databases and have some rudimentary knowledge of the subject matter can be very helpful in the data clean-up efforts

CONSULTANTS

In the Fix It As You Go approach, there can also be a place for specialty consultants. They have expertise and can support the effort in its developmental stages. They can also be used in an outsource model to process changes and return correctly-formatted data for reentry back into the CMMS system. Unlike the Big Bang approach, this group is not large nor do they stay onsite for a long period of time. Their fee structure in this scenario is often transaction-based—a flat fee per record. In addition, specialty consultants can provide specialized software for internal personnel use. This software is purposely designed to extract, cleanse, and then return data to live systems. These applications are preferable to attempting to clean the data directly in the live system.

As you can see, validating and correcting asset and material items data in this fashion takes everyone's involvement over an extended period of time. That extended time duration and the complexity associated with the distributed responsibilities makes success in this mode quite difficult. The likelihood of long-term success has traditionally been low. Companies reorganize, they are sold, people get promoted, and other events take place that serve to break down even the most well-constructed effort. Specialty consultants address this risk by developing and offering fix it as you go models that leverage technology; these models don't depend as much on a myriad of people distributed throughout the organization. Some of the models can easily extract portions of data from live systems, cleanse it to agreed-to standards, and reload it back into the live system.

Costs

The cost of this approach is difficult to determine because a great many people are handling the data integrity tasks on a part-time basis over an extended period. Costs can be estimated, but the ability to actually track part-time costs for a large population of internal resources conducting this work is unlikely. This effort is spread over an extended period of time. Therefore, it is likely that it will be more expensive (assuming that the various people in fact do spend a portion of their time doing the activities) than the Big Bang approach, although the costs are less likely to be noticeable because they are "buried" in other budget line items. In any event, the time to achieve the intended benefits is certainly delayed, negatively impacting the financial justification that may have launched the effort in the first place.

Sustainability

Unlike the Big Bang approach, the Fix-As-You-Go effort is not a project—it has no end. The sustainability issue is really about keeping all those who are involved both aligned and engaged in handling their portion of the overall effort. With all of the changes taking place in organizations today, sustainability is going to be more difficult in the fix it as you go model. These efforts often start off with a great deal of focus, but over time they degrade and lead to a less-than-desired final product.

One solution to the sustainability problem is to focus the clean-up effort on the critical equipment, as identified by the functional experts in the plant. This lessens that workload. When one of these assets is taken out of service for repair, the data is validated and noted in the file. Others that are identified as non-critical do not get the same level of attention. In these cases, maybe all you do is validate the essential data elements.

13.7 The Line in the Sand—More on Sustainability

There is one other aspect of sustainability that is an integral part of a data clean-up effort. It deals primarily with the installation of new assets or those that are part of the replacement process where the replacement is not an in-kind change. In these cases, there needs to be a process in place within the plant that says, "from here on out, there will be no replacement or new asset installed without the inclusion of those involved in the data integrity process." This is the proverbial line in the sand—a set time after which all replacements and new assets must have a data integrity component built into the process. This approach is not difficult to accomplish.

New or replacement assets come into being in two ways. Data integrity solutions need to be part of both processes.

In-Kind Replacement Assets

These assets replace existing assets that have worn out. These new assets may be a replacement-in-kind, but they may have different associated data, such as serial number or manufacturer. In-kind replacement assets are typically installed by the maintenance organization. Associated with this asset replacement will always be some level of planning and scheduling so that the asset is installed as required. Just as the planners have responsibility for printing out the existing asset data sheet, having it updated in the field, and making certain that the corrected data is entered into the databases, they have a similar responsibility for the replacements.

Capital Replacement

In the case of capital replacement, you're addressing a totally new asset that does not exist in any of the current databases. Quite often, engineers are totally focused on getting the installation complete and turning the asset-related data over to the maintenance organization. As stated earlier, the capital project engineers must have responsibility and accountability to assure that the asset-related data is entered into the system. The best way to handle this is to prohibit closing out the project where asset data is involved without signoff by those responsible for asset database maintenance. This process can be easily constructed. It essentially compels the engineers to make certain that the asset-related data is entered into the system. Lack of signoff prohibits them from closing-out their project.

With these two processes in place, you can reasonably assure yourself that replacements-in-kind and installation of new assets will have the correct data added when they are installed.

13.8 Commitment to Doing the Work

Not only is it important to ensure that the asset database is accurate, it is also important to visibly see that the organization is serious about data integrity. Employees are very aware of what is going on in an organization, often because at some level they are involved in the various effort. Regardless of which approach you decide to use, the employees will be watching to see what decisions are made regarding the data clean-up process. If the organization shows a level of commitment, dedicates ample resources, and demonstrates belief in the value of data integrity, the employees will positively respond. This will engage the workforce because they will clearly see that the organization believes in the value of data integrity. Failure to take proper actions with regard to asset-related data clean-up sends an entirely different message to the organization that asset data is not important and, as a result, data integrity will suffer.

PART THREE

Sustaining What You Have Created

If you have taken the time to understand the scope of the problem and made the effort to address it, then you should be realizing the economic value that reliable asset data provides. However, this effort is not a project that has a beginning and an end. True, it has a beginning, but the reality of the situation is that it has no end! Maintaining what you have created through hard work is vitally important because data integrity is dynamic; it is always changing as new equipment is installed, existing equipment modified and old equipment retired. Therefore you need to implement a process to sustain what you have created. If not then all of the time and effort will have been for naught. Part Three addresses this important topic.

Chapter 14

Data Governance

> Without order there is only chaos.
>
> *Unknown*

14.1 Data Governance—Insight to the Problem

In past chapters, we have learned about the asset and asset data life cycles. Consider the following scenarios and see if you can identify what they have in common:

❏ The engineering firm designing your most recent project delivers the asset-related data in a form other than what you specified. You accept it anyway and file it away in Engineering.

❏ The design firm creating your process flow diagrams incorrectly numbered the assets in spite of the fact that you had held several meetings on the subject. Unfortunately, the contract was loosely written and you have no evidence defining the methodology that was to be used for the numbering scheme. The project is under construction and it is too costly to alter the coding. Even though it is different from the coding of other plant assets, you accept it.

❏ You are constructing three similar systems at three of your plants. The engineering concept was to have the same asset identification at each site. The personnel at one of the sites decided that they did not like the coding structure you were using and they changed it at their site.

❏ It was planned that the asset databases at all of your plants would be set up the same in your new computerized maintenance system. One of the plants changed the coding structure, rendering your ability to search for asset data across plants impossible.

❏ The asset database of the maintenance system was set up so that a large group of engineers and maintenance personnel had access and could make alterations as they saw fit. People with good intentions made entries that were not in any standard format. As a result, asset data was not just difficult to find. In many cases, it was impossible.

❏ When errors were found by the maintenance mechanics in the asset-related data, they initially reported the errors. Later they discovered that there was no process in place to make the corrections, so they stopped. The result: data errors were never corrected causing lost time and effort.

❏ Spare parts for newly-acquired equipment were not entered into the computerized maintenance management system because a procedure was never developed. This resulted in a significant loss of revenue when needed parts were not readily available.

❏ Many of the departments who used the same asset data had no confidence in the database in which this data resided. Consequently, they developed or purchased their own applications. The result was that the data in the different systems for the same data element was often different.

If you have not figured out what all of these examples have in common, the answer is that the organizations that experienced these problems had no process in place to prevent them from taking place. The result was a lack of control over a very valuable part of the business—its asset-related data. This lack of data control caused erroneous analysis, less than optimum decision-making, and most likely a major loss of company revenue as a result.

It is one thing for a company to recognize the importance associated with the control of asset-related data. It is entirely another for it to understand the commitment that is required in order to achieve asset data integrity. Even understanding this commitment is not enough; the organization must be clear about what steps must be taken and how they should be taken in order to attain a value-added process.

14.2 Shifting the Burden

In his book, The Fifth Discipline, Peter Senge identifies a management principle that sheds light on what is required for a value-added data integrity process. Senge refers to an archetype, a model which patterns various work process issues. In Senge's archetype called Shifting the Burden, he provides insight into the data integrity problem that exists within each of the examples cited at the beginning of this chapter. In describing this archetype, he writes,

An underlying problem generates symptoms that demand attention. However, the underlying problem is difficult for people to address, either because it is obscure or costly to confront. So people shift the burden to other solutions—well intentioned easy fixes….

The problem is that the easy fixes do not solve the root cause of the problem. To make matters worse, the root cause often gets worse because it goes untreated. Meanwhile, the short-term fixes appear to have the problem solved, at least for the moment. A diagram of this archetype is shown in Figure 14-1

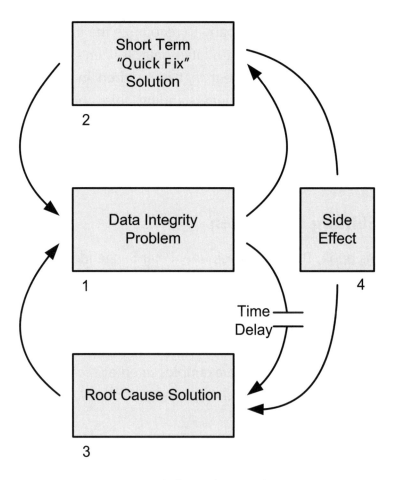

Figure 14-1 Shifting the Burden

(Adapted from Peter Senge, The Fifth Discipline)

Let us examine how this model applies to our problem. The data integrity problem identified in Block 1 indicates that asset-related data is either not available or has questionable accuracy. Nevertheless, data is needed for analysis and highly important business decisions. This need leaves us with a serious problem. The short-term solution is that if our data in the company system is of little value, let's create our own system where we will store what is needed and have a high degree of confidence that, when needed, we can provide it. As a result, personal asset databases are created (Block 2) and data that is of use is stored in this database. Over the short term, those involved are able to provide for the needs of the group.

The result is that the problem of Block #1 seems to go away—but does it? Not really, because personal or departmental databases are loaded with problems, including that data is not universally available and often not kept up to date. The side effect (Block 4) of this short-term solution is that the organization does not address the root cause of the problem.

Recognizing that the short term solution is incorrect and finally addressing the root cause may take a long time. In fact, these steps may never happen as groups continually handle the asset data integrity problem by masking them with quick fix solutions. If the true root cause of the problem (Block 3) does finally get addressed, it is highly likely that it will have gotten worse—now you will have to undo and correct all of the additional problems imposed by the failed quick fix solutions.

14.3 The Long Term Solution

A long-term solution avoids the quick fix and all of the related problems that the quick fix approach brings to the table. However, adopting and implementing this approach is not easy. It requires time and resources—both short-and long-term—and further requires the organization change their data integrity culture: how they think and act about this very important topic.

What we need to put in place is a corporate-wide data governance process. The concept of data governance is drawing a great deal of interest in relation to overall corporate data management, not just for asset-related data. Data governance applies to all data generated, stored, and used by a corporation in the conducting of the business. However, this extremely broad topic, often referred to as master data management, is outside of our scope. What we want to discuss is data governance as it relates to asset data integrity. It is possible for our topic to stand-alone because, as we have seen, it is incredibly important to the many organizations that deal with asset-related data. However, we can only hope that the same problems that are associated with our sub-topic are recognized at the corporate level and addressed globally for all data.

When the term data governance is brought into the conversation, the first question is often about what it is. Data governance, as defined by the Data Governance Institute, is:

> *A system of decision rights and accountabilities for information-related processes, executed according to agreed upon models which describe who can take what actions with what information, and when, under what circumstances, using what methods.*

This definition has close ties to some of the discussions we have had in others chapters.

In Chapter 5, we learned about the asset transform and the three components: Entry, Storage, and Retrieval. If we wish to provide complete and accurate data to the decision-makers of the corporation, then the data governance process needs to be able to govern all three parts of the transform. We know that different departments have different levels of involvement throughout the transform. Therefore, we can quickly recognize that a data governance process must have the ability to control what these departments do with relation to the data.

Eight Elements of Change

There are also a set of strategic elements that must be considered when you think about what is required to place a strong data governance process in place. These are referred to as the Eight Elements of Change; they have great relevance to the data integrity effort.

❑ **Leadership.** Because data governance needs to originate at a corporate level, leadership provides the only way we can assure rigid consistency within a multi-plant environment. Additionally we know that the organization focuses on the areas that leadership considers important. If our governance process is not initiated and managed at a senior level, its longevity and importance will suffer. Leadership is also required at the site level because that is where the data is most often used in performance of the work. Therefore, the organization should not have just a single senior leader endorsing data governance, but a multi-level team making certain that the importance of data integrity is recognized and addressed.

❑ **Work Process.** Any activity that involves multiple departments performing work in a complex manner requires a work process. In the case of data integrity, this is most certainly the case. In order for the transform to work at its optimum, a detailed work process is required. Additionally, because we want all of our plant sites to handle data integrity in the same manner, this process must be global—one that can only be implemented and maintained at the corporate level. We all know what can happen to a process that is maintained at the site level. Each site determines how it will be utilized. Then, over time, any consistency that existed when the process was created disappears. For the data integrity process, this must be avoided at all costs.

❏ **Structure.** It is certainly not sufficient to have a senior leadership data governance structure because we also need structure at the plant and even departmental levels. The problem is one that often emerges in organizations that have decentralized control in which the site mangers have a high degree of autonomy. In these cases, where does the role of data governance fit? If it is a defined position, does it report to the site or to corporate? Which reporting relationship is dotted line and which is solid?

These questions are not easy to answer; they differ site to site. However, the thing to recognize related to structure is that 1) you need to have overall management of data integrity at the corporate level and 2) you also need management at the site. One way to blend this into a single solution is to make the data integrity process one of the key performance measures of the site manager. Then, have the manager's performance reviewed by corporate data governance. This approach to structure can solve the problems we have just discussed.

❏ **Group Learning.** This topic also requires a centralized governance process. Continuous improvement is a trait that all efforts of this nature aspire to achieve. Because the process is never the best it can be; we constantly learn better ways to accomplish work.

Without group learning in the mix, we will never capture this very important aspect of data governance. Without learning and continuous improvement, we still would be in the paper and filing business as opposed to handling much of our asset-related data electronically. Furthermore, without a level of centralization, learning would never be shared and consistency would suffer.

There is another important aspect that is closely tied to the element of group learning. That aspect deals with how we learn and what we do with what we learn. How do you learn about success and areas that require improvement within the realm of asset data integrity? You learn by auditing the process and the controls over the process imposed by the data governance process. If the process is working, you learn that what you have put in place is properly serving the needs of the business. If things are not working, as uncovered by the audit, corrective action can be taken to bring the process back into alignment. Failure to learn what works and what doesn't work and then fix the problem areas is a central concept for sustainability of this very important part of the business.

❑ **Technology.** In the world of data governance, technology is both a tremendous benefit and a danger. Technology in general and how it handles data integrity at all levels of the data integrity transform is improving constantly. It is important that those involved with data governance understand the changes and take advantage of them as appropriate for the organization. That can be the benefit. On the other side of the equation is the danger. In data integrity systems where the correct data is often not available, good people with good intentions will create or purchase databases of their own to solve their short-term problem. Left unchecked this becomes an enormous problem when corrective action is finally taken.

Therefore, not only does data governance need to help the organization stay current with the technology, but also the governance process must protect the organization from the lurking dangers.

❑ **Communication.** As we have noted many times, data integrity is not high on the list of topics that people discuss, that is, until they need the data and cannot find it. It is the job of the data governance process to change that focus and to make people aware of the importance of data integrity through the communication channels within the company. This effort also sends a message that the leadership considers data integrity an important area and as we know, the organization typically will focus their attention on what the leadership considers and communicates as being important. If data integrity is to deliver the value that is inherent in the process, then communication cannot be a one-time event.

❑ **Interrelationships.** From all of our discussion, it should be clear that the data integrity effort affects everyone. For the process to work at its highest level, positive interrelationships across inter- and intra-organizational boundaries are critical. It is the responsibility of the data governance effort to assure that these relationships stay positive and that issues are proactively addressed and resolved in a timely fashion.

❑ **Rewards.** The most effective rewards are those where the action and the reinforcing reward are closely tied in time. In other words, you perform an action and the associated reward takes place shortly thereafter. Unfortunately, data integrity does not have this component. In most cases, asset data is added to the system by one group and not retrieved until much later by another. Where is the reward for the group that properly put the data into the system? There usually is not any. In fact, they may never know that the data they so diligently entered supported analysis that saved the company money.

As difficult as it may be, this barrier needs to be addressed by the data governance process. Otherwise, people will begin to question the value for the work that they are doing and focus their work efforts in other areas. Continuous communication with real-life examples can serve to bridge this problem. It allows those working within the three stages of the transform to see the value that they create, even through there is a time delay.

14.4 The Benefits of Data Governance

We have not yet discussed the various roles in the data governance effort. Yet reading between the lines clearly indicates that there will be roles both at the corporate and site levels and somehow—solid or dotted line—they will be linked. It is not easy convincing the leadership team that this role is critical to their business because the short-term fix often clouds their sight. Just as we need to create a business case for data integrity in general, we also need to create one for the data governance effort and the often full-time jobs that are included.

Why is this needed? Although the huge savings associated with data integrity will get management's attention, the next question that will be asked is why this work cannot be done as a part-time effort included within the job description of those affected. In some cases, the answer is that it can be a part-time effort. However; in other cases, the work requires a full-time effort and additional resources to make the data integrity / data governance process function. Even some of the part-time roles require full-time commitment as the governance effort is established and can only revert to part-time after it is firmly in place.

Therefore, to obtain this resource commitment, it is critical to explain the benefits of having a strong data governance effort and the downside of not having one in place.

At the senior leadership level, data governance oversight can be a part-time effort included within the job description of one of the team members. However, at the outset of a data governance effort, this is not the case. At the beginning, direction from senior leadership is a full-time effort for someone from the leadership team. This assignment provides the focus needed to entrench the process in the business culture. Over time the level of importance imparted to that person will go a long way in helping to gain site support and buy-in. As we progress lower in the organization's structure, some jobs are definitely not part time, but require the full-time attention from those individuals assigned.

So how do you convince the leadership team that they need to provide both part-time and full-time resources in an area that did not require them prior to your discussion? The business case (Chapter 1) should most certainly have gotten people's attention. The next step is to provide the benefits associated with success, then convince the team that these benefits cannot be accrued without people focused on this part of the business. The benefits from having a data governance process and the associated resources include:

❏ **Oversight.** Without governance resources, whatever process is put in place will degrade over time. The process needs a central figure who can dictate how it is supposed to work and make the needed adjustments when it is not working as planned. Full-time oversight at the outset is very important, especially in multi-plant operations. In these situations, the autonomy of the individual sites often gets in the way of consistency across the entire corporation. Therefore,

any members of the senior leadership team need to have accountability and very detailed involvement at the beginning. They need
to set the stage for "how things are going to be" and have the
authority to require no deviation.

Part-time efforts in this area set the stage for failure before the
effort is even off the ground. As time goes on, this full-time role can
be altered to part time. This alteration may not happen for a considerable period of time; until the senior team is firmly convinced that
data integrity is an integral part of the organization's culture.

❏ **Policy and Control.** Along with oversight come policies and controls. Policies put into place work processes. In turn, controls
enable the policy to be executed on a continuous basis. Setting policy and putting controls into place are further responsibilities of the
individuals charged with oversight. However, they cannot handle
this task in a vacuum. They need input from people in the field who
are closer to the real work. These employees can help make the
policies and controls something that will work when deployed.

As with oversight, those involved with creating policy and controls will need full-time involvement, at least until everything is
developed and deployed. Unlike oversight, however, this full-time
effort should not take as long. Why? The reasoning is that controls
cannot be developed over an extended period. Instead, they need
to be developed as soon as it is decided that a robust data process
is going to be put into place.

❏ **Delay.** in development and deployment can be catastrophic. The
organization needs immediate guidance and this is what the controls will provide. Without them, the organization—although striv-
ing to comply with the data integrity direction of the leadership
team—will create policy and controls of their own. This is bad in

two respects. First, if you have multiple sites, the answer developed at each may be very different. Second, once these controls are in place, they take on a level of ownership and are hard to dislodge.

❑ **Monitor / Audit.** Even the most effective process can go astray. The benefit of a data governance function is that the process is continually monitored. Specific areas where problems appear to exist are audited. As a result, you learn rather quickly if something is going wrong and can set in motion a process not only to identify the problem but also to fix it as well. The function is not a part-time effort; it requires full-time data governance involvement. As we shall see, this function is not the only one that the full-time personnel can provide. However, it is an extremely important one.

Failure to monitor and audit properly can expose the data integrity process to serious problems. These include lack of attention, not following the rules, new employees without training, and other tasks that management (especially if they are reactive) believe are more important. Consequently the work gets re-directed away from data integrity.

❑ **Corrective Action.** There is a benefit to the monitoring and auditing process. If you have a strong process, then the members of this team will uncover problems that could derail the data integrity effort. Corrective action is what keeps the train on the track. A secondary benefit of corrective action is process reinforcement. As people submit data problems and see that action is being taken to correct them, they will begin to understand the level of importance that management places on the data integrity effort and will work to support it.

❑ **Assigned Responsibility.** Having dedicated resources assigned to the data governance process will assure that all phases of the data integrity transform are functioning properly. The jobs associated with data integrity governance will be described in the next section. Initially there may be a need for more full-time resources than will be needed after the data integrity effort is stabilized. The reason is that initially there will be a great deal of corrective action and data clean-up in order to attain the level of desired accuracy.

Collectively each of the above benefits provides the organization with a data integrity process that will be sustainable. The process will have a low likelihood of degrading to the point where it is not viable. Without resources to fill the positions requiring full-time or part-time involvement, these benefits will never be accrued, leading to many if not all of the problems we have described to this point in the book.

14.5 The Jobs of Data Governance

As we have discussed in relationship to the RACI chart provided on page 268, there are several levels of jobs within the data governance process.

❏ **Executive Level.** This position is the focal point for accountability within the process. The fact that a key person on the leadership team has been assigned this role should send a clear message to the entire organization about the importance the company attributes to data integrity. Once the process has been ingrained in the business culture, this leadership role can be one of many assigned. However, this is not the case at the outset. In the beginning, the executive position has full time responsibility as oversight teams at the various plant sites are established, policies and procedures are developed and deployed, and auditing is conducted to assure everyone that the process is indeed working as planned.

❏ **Oversight Teams.** This team is composed of key members from corporate such as finance, safety, and environmental, along with their counterparts at the various plant sites. Obviously, the more plant sites that exist within the company, the more oversight teams that will need to exist. There needs to be a team at each site to assure that the process is properly functioning. In addition, elected site personnel will work on a corporate team chaired by the executive sponsor. Having this multi-level structure will enable consistent practices to be deployed across the business.

❏ **Plant Departmental Role.** The departmental members of the site oversight team also have another role. It is their job to bring the data integrity policies and procedures to their individual organizations, make sure that the requirements are understood, and that the effort within the department follows the established guidelines. In a sense, they have a dual role. Upward they are part of the site team working strategically at the site. Downward they are responsible for the data integrity actions of their individual department—accountable to the site's oversight group.

❏ **Specialists.** An immense amount of work is required to 1) clean up the existing process and 2) keep it clean as changes and / or additions take place. This is a full-time job for a group of data integrity specialists who have been trained and know how to make the process work. Having full-time personnel will assure the organization that there will be proper controls and consistency in place. It will also enable the organization to clean up the "sins of the past" and get the data in the correct format.

14.6 It's All About Policy and Controls

Of all of the aspects of the data governance process, one of the most important is policy— and how the policy is controlled. The senior leadership and oversight teams are responsible for setting policy and overseeing the process, assuring that the controls are in place to enable the policy to function. All of the other aspects of the process are then driven by the policy so that the process functions as designed. Therefore, it is important to provide the framework of what goes into an asset data integrity policy. The "how" part is left to the reader because all companies and plant sites are somewhat different. Knowing the components will allow you to create the policies for asset data governance in a manner that will fit your company and its resident culture.

Gwen Thomas, president of the Data Governance Institute, has described what needs to be included in a data governance policy as CRUD. Yes "crud," but not in the sense you would immediately consider. In our context, CRUD represents the various ways in which we handle our asset data. Understanding these aspects enables us to build policy to address these four critical areas. Therefore, CRUD stands for:

- ❑ **Create (C).** The ability to create asset-related data
- ❑ **Read (R).** The permission level that allows data to be read
- ❑ **Update (U).** The capability to update data—either new data or data that has changed
- ❑ **Delete (D).** The ability to delete asset-related data from the system

Because each of these initials carries with it a whole host of policy opportunities, we need to address each so that, as you build your policies, you address all of the components. After all, once the policies are developed, they will be rolled out and utilized. Therefore, be certain you have a chance to think about what CRUD really means to your organization and your asset data integrity process.

❏ **Create.** In the portion of the policy related to creating data, you need to address all the various ways that new data gets into the system. This is the entry part of the data integrity transform. Identify all of the sources of new data, where they come from, how what is delivered will conform to your company's data integrity delivery specifications, how the data is indexed, and most important who has the responsibility for the actual creation of the records in the system.

❏ **Read.** It is not true that everyone should be able to read all asset-related documentation. Therefore, the policy needs to establish levels and controls for who is granted access to the various "read" levels. Suppose that the data integrity storage system retained corrosion and thickness reading for all of your major fixed assets. Furthermore, one of the assets was experiencing thinning due to heavy corrosion. Although the plant maintenance organization is addressing the problem, there may be individuals who should not have access to this data. Otherwise, erroneous conclusions regarding the safety of the plant could be drawn. Other aspects of the Read portion of CRUD that need to be addressed include who can read the data and what they can do with it. For example, you may want to allow someone to read the data, but not allow them to copy it and use it elsewhere within the business.

❏ **Update.** Some organizations believe that all individuals in middle and higher management (or a vast majority) should have the ability to update asset-related data. As a blanket statement, this is a major mistake. If given the opportunity, there are those who with good intention will update the data. The problem is that they may not do it frequently. In turn, they may not follow the correct data integrity conventions or they may simply make mistakes because they do not know the conventions.

In general, a policy needs to be enacted that severely restricts those who can update the asset records to those trained in how to update them properly. There may be some applications where selective data element permission can be granted to a larger audience, allowing for selective updates. However, even these situation need to be controlled. Furthermore, an audit trail within the system needs to be in place before this level of permission is enacted.

❏ **Delete.**   Asset-related data should almost never be deleted. Archived, yes, but not deleted, which implies it is erased and not recoverable. In fact, there are computerized maintenance management systems that allow assets to be made inactive, but not deleted. Most companies do allow deletion if an asset is totally removed from the plant. However, these companies also impose what is called a data retention period—although the data can be deleted, this action cannot be taken until several years after complete removal. After the retention period, the policy should dictate that deletion is required in order to conform to both the data integrity policy as well as the legally imposed retention policy.

The CRUD concept is an important part of the policies and procedures that will be developed to support the data integrity process.

Make sure that the details are given sufficient attention; these are what you are going to audit when you validate that the process is functioning properly.

14.7 Roles and Responsibilities

Everyone has a role within the data governance process. Because of the numerous roles and large number of involved groups, the complexity of the roles and responsibilities are not easily portrayed. One way to do this is with a RACI chart, which describes the various levels of involvement with the data integrity process. RACI stands for the four levels of involvement that people can have with any task: Responsible, Accountable, Consulted, and Informed. To show this involvement and reduce the complexity of the effort, the RACI chart was created. The various aspects of the data integrity effort— from overall governance to those who work full-time for the "care and feeding" of the effort—are listed vertically. We list the roles that are involved horizontally. Within this arrangement's matrix structure, we list the responsibility (R, A, C, I) for each role. The definitions of the four terms in the RACI chart are as follows:

R—Responsible: Individuals assigned an R participate in the completion of a task and add their expertise as required. More than one role can be assigned an R for a specific task.

A—Accountable: The people assigned an A have sole accountability for completing the task. However, the task can be completed by one or more individuals assigned by the accountable person. Those assigned an A are usually managers who then delegate the work to those assigned an R. Nevertheless, the success or failure of the task rests solely on the shoulders of the A.

C—Consulted: These individuals must be consulted and approve prior to the task being completed. There can be more than one C for a specific task. Unfortunately the more people designated C, the harder it may be to get approval to proceed. They all need to be consulted and approve the work.

I—Informed: These individuals are informed of the progress and next steps in completing the activity or task. There can be more than one I per task.

On the following page, in Table 14-1 we provide a RACI chart for the data integrity effort. Everyone is included, from the senior leadership level where ultimate accountability is assigned all the way through the organization to and including the workforce, who at the minimum will support the effort by identifying data integrity related problems.

The rationale of the RACI chart may be clear to the reader. Nevertheless, let's consider two of the elements and show how the chart functions.

❏ **Structure Provided and Maintained**—Once the overall process has been established, an active supporting structure populated with both full- and part-time employees is needed. This task falls to the oversight team to develop and put into place (A). They have been given the assignment by the senior leadership team and are the ones who will need to identify and make available the resources required. They have the responsibility (R) to make it happen. As this effort progresses, the executive data governor has a consulting / approval (C) role working with the team. They are Cs because they do not actually do the work, but they have final approval before

Table 14-1 Data Governance RACI Chart

	Executive Level	Oversight Team	Plant Departments	Data Specialists	External (Supplier / Vendor)	Workforce
Data integrity is established as high priority	A,R	C	I	I	I	—
Continuous improvement is in place (Learning)	A,R	R	R	R	R	—
Processes and procedures are developed and put in place	A	R	R	C	I	—
Structure is provided and maintained (full- part-time)	C	A,R	R	I	I	—
Communication is ongoing to support the effort	C	S	R	I	I	—
Process administration is in place to assure compliance	C	A	R	I	I	—
Perform detailed data integrity tasks	I	I	A,R	R	R	—
Process is measured and audited	I	A,R	R	I	I	—
Corrective action is taken where problems are uncovered.	I	A,R	R	R	R	I
Technical aspects of the effort are addressed	I	A	R	C	C	—
Problems are identified for corrective action	—	—	R	R	R	R

implementation and assignment. The data specialists and those external to the organization need to be informed (I) so that, as required, they can support the structure.

❏ **Corrective Action**—As the data integrity process is monitored, problems will be identified. The responsibility to identify the problems and participate in their solution rests with the oversight team which makes the team both accountable (A) and responsible (R). Of course, they need the individual departments to perform work on the corrective action so that these departments also have responsibility (R). Because corrective action, depending on the problem, could require work by the specialists or external organizations, they too are assigned responsibility (R). The senior executives for data integrity need to be informed (I) so that they can feel comfortable that process issues are being resolved. Depending on the result of the problem solving process and the required corrective action, these individuals may also be consulted (C) before implementation.

The other elements have the same logic. It is important to remember that there can be only one role assigned an A whereas other roles can have multiple assignments.

14.8 When Should We Start?

Thinking and doing something about asset-related data integrity, especially if your current process either does not exist or is critically broken, is a daunting task. It can be so daunting that you lapse into a state of paralysis and nothing gets started. Do not wait; start now! Assign the executive level person, develop your policies and procedures, and make today the first day of the improved effort. Once you have everything in place to move forward under controlled conditions, only then figure out what needs to be done to clean up the problems of the past. This latter effort is going to take a considerable amount of time and effort. In the end, one good business decision driven by accurate asset-related data will easily pay for all of the hard work.

Chapter 15

Sustaining What Has Been Created

> Without a sustainability effort the best day for the data
> will be day one,
> and it will be downhill from there.
>
> *Steve Thomas*

15.1 The Need to Sustain

You've gone through the processes of learning about the importance of asset-related data, cleaning up what you have in place, and establishing a data governance process. Now it is absolutely critical that you have a plan in place to sustain what you have achieved over time. If you don't seriously consider the sustainability aspect of this work, then save yourself some time, money, and frustration—and don't even begin!

The business landscape is continually changing. Companies merge, people move into and out of jobs, people leave the company and new ones enter, and—believe it or not—senior management's focus on what is important changes as well. Data integrity is serious business. In spite of all of these changes, we can't allow the effort to degrade, not even a little. The data governance process that was defined in Chapter 15 is a good start, but it has to be tightly focused on sustaining what was put into place.

15.2 Establishing Ownership

The key to maintaining data integrity is ownership. In this case, we are talking about specific owners of specific parts of the data set. It is not enough to say that the Planning Department has responsibility for the integrity of equipment records or that the Procurement Group is responsible for the integrity of the material item records. Ownership of the data sets, or portions of the data sets, should be assigned to individuals within an organization as a part of their job responsibilities. However, this isn't as easy as it may appear. Often there are multiple users and consumers of the data and the designated owner of the asset may not be one of them.

Consider a pump in a process plant. The production organization operates it, maintenance repairs it, reliability engineers work to make it more efficient, capital projects investigates how to upgrade it, and the safety and environmental organizations are concerned about possible risks to the plant and employees. In this example, production would be the primary owner of the asset because the pump is critical to the operation of the manufacturing process. However, the primary ownership of the data resides with maintenance and reliability and, to a lesser extent, the other organizations mentioned.

This approach makes sense because the various departments identified have the most to lose if the data is flawed. This philosophy also fits with the governance process. However, it is insufficient to assign responsibility without authority. Data owners need to be empowered and feel empowered to control the health of the data. Management must provide this empowerment. They must publicly and openly appoint the data owners while stressing the importance of data integrity. In addition, processes must be in place that further empower the owners to act on and correct bad data.

15.3 Communication

Communication plays a critical role in the sustainability process. Once management has appointed and empowered the data owners, the next step is establishing a facility mind-set that focuses on the importance of data integrity. Educating the organization on the role of quality data in both the short-term management of the facility and the long-term strategic planning necessary to sustain the business model is critical to success. Emphasizing the role that data plays in all of the organization's decision making processes is vital. Simply put, the message is that bad data leads to bad decisions; good data leads to good decisions. For many organizations, it is a new idea that data—especially asset and material item data—is so important.

Although production and operations organizations have been data-driven for years, the concept that such hard data is vital to the entire organization may come as a surprise. Culturally, most organizations have not thought about this. To reinforce this mindset, and to assist the organization in understanding and embracing the importance of data integrity, there is a need for the senior leadership team to consistently and constantly provide this message. This can be accomplished by providing examples of how accurate asset-related data supports sound business decisions. Simply telling an organization that data is important is not sufficient. Changing how an organization thinks about data is the key.

15.4 Process and Procedures

We have now identified the data owners and established the importance of data across the organization. Next, we must create the tools and processes that will help sustain the model. This process begins with developing detailed flow diagrams that depict how data changes in the system—the types of changes and the controls necessary to ensure that the changes are recognized and captured by the system. Begin with all the ways that static data can change. Identify where in the organization these changes originate, who may create the changes, and how frequently the changes may occur.

Recognize that changes to your data vary in size, complexity, and importance. The processes you are designing are intended to capture the changes to your data sets, but a "one size fits all" approach is not likely to be successful. A business process that is too complex may be ignored for minor changes, for example, the repair part being installed from a different manufacturer. A process that is too simplistic may not provide the level of detail needed to capture all the required information for a large capital project. A balance must be struck between a level of complexity that ensures the consistent capture of data and a level of simplicity that encourages compliance. Finding this balance may often result in the creation of several processes—or a single process with multiple paths—for the different types of changes identified.

Remember that it is important that all changes to the data be captured, no matter how big or small. Generally speaking, there are relatively few big changes (capital projects, plant betterment initiatives), but a very large number of small asset or material item changes (equipment replacements, parts substitutions). It does not take very long for a data set to become corrupted. Once corrupted, for whatever reason, the confidence of the user community is eroded. Therefore, as you develop the process model, it is imperative that the model be tested for all potential ways change can be introduced into the data set.

15.5 Training

The next crucial part in sustaining data integrity is training. Keep in mind that much of the staff has traditionally not thought of data management as an important part of their jobs. For some, data is just a necessary evil. Therefore, when developing training for the new processes and procedures for data management, you must address the "why" of data management, not just the "how." The training must make clear why data integrity is important to the organization, the business unit, and to them personally. The last element may be the most important.

Many studies have shown that unless the question of "What's in it for me?" is answered, the ability to sustain change is significantly diminished. In fact, studies have shown that unless the cultural change aspect of any large project or organizational change is addressed, the chances of successful implementation are significantly diminished. When addressing the "why" of data management, link the importance of data to the organization's mission and business objectives. Then create an understanding of how data drives business decisions. It is vital that everyone comes away from the training believing that good data is key to good decisions.

Training should be targeted to the individual users' specific needs.

There is a vast difference between how each department relates to and affects the data. For example, an engineer who typically works with new assets is involved in a very different manner than a maintenance planner or operator. Another important aspect of training is to provide everyone with an overview of the entire effort so that they can clearly see how their efforts fit in the big picture. Without this aspect provided in the training, we are reinforcing the silo mentality—the very thing we wish to avoid.

When developing training, start by assessing the training needs of the different constituencies in the organization. For each group, ask "What is the student expected to know at the end of the training?" For some, the training will need to include the software that is being used. For others, a focus on the process steps may be all that is required. To be most effective, all training needs to be hands-on and realistic for the end user.

15.6 Prepare for Data Growth

In today's business world, the amount of data collected and maintained is increasing at an astounding rate. Recognizing this phenomenon, there are two very important checks that must be made once you have established your basic process model for data integrity. The first is to take a long view into the future. The data model needs to be examined to determine if it can handle data growth. Can the business model expand to accept more users of the data or more personnel involved in collecting, reviewing, or integrating data into data sets? What will happen if leadership or ownership for the data changes either through normal business growth or attrition? Can your network expand to meet the storage requirements of the ever-increasing amount of data to be housed? If the business process was developed for a single facility, can it be expanded across the enterprise without significant modification?

Although some of these questions may be unknown or unknowable as you build your business process, these questions need to be asked and answered as best as possible. With the amount of time and money that has been invested to get you this far, it would be devastating to outgrow your business model in one, two, or five years because growth was not considered at this stage.

Another aspect of data growth that must be considered is the processing capability of your system. As the amount of data in the database grows and the number of users of the data increases, what will happen to the response time of the software? You must consider this aspect in light of the potential for data growth. If the response time of the software slows as data increases, you will likely see "work-arounds" as the user community becomes frustrated with the system. Work-arounds, or working at all outside the established processes, can quickly lead to loss of data integrity.

15.7 Walking the Walk

Once the data governance, sustainability process, and training have been completed, it is time to make data integrity part of the organization's culture or "here is how work is done around here." This process begins with the formal endorsement of the concept at the highest level of the organization on down to the workers in the field. As stated earlier, changing how an organization thinks about the importance of data is the key. Senior staff members must not only speak about the importance of data, but also act on it accordingly.

Senior staff should never short cut the newly-established processes or let anyone else do so in the name of expediency. Working in the new processes may take longer until the organization becomes totally familiar with how it works. It would be easy, in times of crisis, for management to be impatient and want the work done more quickly. People may resort to "the old ways" under the perceived pressure from their superiors. But just one such incident can send a crippling message that the data integrity was not as important as expressed. Senior staff must be ever vigilant not only to talk the talk but also to walk the walk.

15.8 Quality Control and Quality Assurance

Once the organization is operating utilizing the data integrity model, the job is not over. A robust Quality Control and Quality Assurance process must be in place to ensure data integrity for the long term. Let's make a distinction here between Quality Control and Quality Assurance. For our purposes, Quality Control is the continuous and ongoing data integrity checking performed by the governance organization and the data owners. Quality Assurance is the independent review to ensure that established business processes are being followed and that those processes continue to deliver the required results.

First, let's discuss Quality Control (QC). Your established business processes should have QC checks built in at appropriate points. These start as early as the data building phase of a project. They include peer checks of data entry and supervisor reviews. QC continues throughout the life of the data with in-process checks when data is revised or added. These checks are no different than measuring the dimensions of every so many widgets as they come off the end of the assembly line. The widget is checked to ensure that the critical dimensions are within tolerance. The number of faulty widgets is recorded and the percent of deviation is noted. In this way, a statistical model can be developed for the "normal" error rate for widget production.

The same principles can be applied to data integrity checks. Errors in certain data fields are inconsequential. An error in other fields leads to useless data. Although data integrity is important to the organization, the cost of have zero defects in the data may be too overwhelming. Establishing the importance of each data element allows you to focus your QC efforts where they do the most good. Just as in the example of the widget, not every dimension is of equal importance. Some dimensions are critical and have very tight tolerances. Other dimensions are less critical and have a wider tolerance.

By instituting a robust QC process for your data and understanding the importance of each data element, you can establish a statistical model that will enable you to sustain a level of data integrity that meets the business needs of the organization. When data errors are uncovered, they should be reported in your Corrective Action system so that correction can be tracked to completion. Remember, data defects can be counted, tracked, analyzed, and corrected just like product defects.

Quality Assurance audits of your data process are also an important element to help you maintain what you have built. QA audits should focus on the processes and procedures used to develop and maintain your data—not necessarily on the data itself.

QA audits should be performed by personnel familiar with auditing processes. Like any good audit, there should be an audit plan established ahead of time and the plan should be followed. It is not necessary to audit all aspects of the data system at one time. In general, it is better to deploy shorter focused audits of discrete aspects of the data processes and procedures rather than large general audits of the entire process. The shorter, focused audits ensure there is sufficient time to drill down and get the important information to the surface. Due to time constraints, large generalized audits are likely to gloss over some critical elements.

While auditing the processes and procedures, the auditor should focus on three key areas. These are the organization's compliance to the process, ease of use of the process, and consistency of the process results. When looking at compliance, you must look to see that all anticipated users of the process are complying and what level of compliance they are attaining. Failure of a significant part of the organization to use the procedures, or a tendency to use the procedures only partially, should lead you to examine the ease of use. It will be necessary to determine if the lack of utilization is due to difficulty of use or mere complacency. The third element is to determine if the processes and procedures are generating the results that were intended. Even with a high degree of compliance, if the results are not what were intended, the problem lies with the structure of the processes themselves.

The QA activities associated with your data integrity are also an important part of any continuous improvement effort. Findings from QA audits should become opportunities for the organization to evaluate if the processes and procedures are continuing to support the organization's data needs now and into the future. Here's another method for using the QA process to drive continuous improvement. Make it mandatory that organizations that are to be audited perform a self-audit prior to the official audit. In this way, the organization can recognize issues and be prepared to discuss solutions when the audit occurs. This technique should also invoke discussion on how the processes can be improved in other areas besides those identified either in the self-audit or audit process.

Again, as in the discussion on QC of the data, all findings should be entered into a corrective action tracking system. This ensures that findings are driven to closure. As with any QA audit process, any findings indicating that established data management processes and procedures are not being followed should be considered as high priority. The time to respond to high priority findings should be short. The longer it takes to address and correct procedural non-compliances, the more likely that severe data corruption may occur. The recovery from such an event can be extremely time-consuming and costly. Repeat findings should also be a major concern. If non-compliance has occurred after a corrective action was instituted, a new high priority finding should be issued, again with a short response time. Repeat findings are also a reason to increase audit frequency.

15.9 Using Key Performance Indicators

In a further effort to help the organization understand the importance of maintaining data integrity, it should establish, track, and publish Key Performance Indicators (KPIs) associated with data integrity. Establishing a target for data defect rate is a simple example. As discussed in the section on Quality Control, it is possible to establish a statistical model of defects in the data. A norm can then be established for data defects. Through the QC process, the actual value can be compared to the norm on a regular periodic basis. Management can analyze the trends of the data and determine if the deviation remains within a statistical upper and lower control band. Data outside the control band will require management action. Again, using the corrective action program to document and track actions associated with the deviation drives the action to completion.

Another important aspect of using Key Performance Indicators is making them available for all to see. Including the KPIs associated with data integrity with the other top indicators that are routinely tracked demonstrates the importance the organization places on data integrity. Many organizations already include satisfactory performance against top level metrics in their variable compensation packages. Isn't data important as well? Getting the data into a viable condition and maintaining it that way has likely been a costly journey. Unless the organization feels some level of ownership in the maintenance of that data, the success of your data integrity program is severely jeopardized. The KPIs are important enough that they should be discussed in staff meetings and morning tool box meetings. However, if you want the KPIs to be meaningful to the people you are sharing them with, it is important that these same people can actually influence the KPI.

15.10 The Continuous Improvement Cycle

A continuous improvement plan must also be maintained for your data and data-related processes. Your business changes over time; so should your data maintenance program.

Periodic meetings should focus on what parts of the data management processes are working and what parts are not. If the processes are no longer working, you need to analyze why. The reasons for the process no longer serving the business needs can be many—they can be minor or very major. Some of the minor reasons can include the introduction of different software applications or upgrades to existing software, requiring modification to the interface points. A new regulation could have been issued which would require additional data or different data to be collected to demonstrate compliance. These types of changes are relatively minor and generally don't require significant revision to the high level business processes. Still, they need to be addressed.

The review meetings with your data owners, users, and data suppliers should be scheduled on a regular frequency, perhaps quarterly or semi-annually. Once scheduled, these meetings should not be skipped. It is sometimes too easy to say that all is well and not ask the tough questions. By adhering to the meeting schedule and asking the questions, you afford yourself the opportunity to correct small problems before they become big ones. These meetings become an ideal opportunity for your core team to review the KPIs you are collecting. Not only should the data trends be reviewed, but also the continued validity of the KPIs should be questioned. Results of any QA audits performed since the last meeting should also be reviewed at this time. If the core team responsible for your data integrity had not been involved with the specific audits, reviewing the findings provides them with an opportunity to identify similar areas for improvement within their circle of influence.

Sometimes there are major changes within the business environment that should trigger an immediate review of your data integrity processes. When the changes include acquisitions or mergers, you must address whether or not your data integrity process will work in the new larger organization. If the existing business process will work, how will data from the new organizational elements be assimilated into the current structure? If the major change is associated with a divestiture, other questions need to be asked. What parts of the legacy data should the organization retain, why and for how long? The last of the major changes to be considered is changes in organization structure or organizational direction at the higher level of the organization. Changes such as these often require that the organization's leaders receive different information.

After determining if there is a need for different information to be conveyed, you must assess if the current processes are capable of supplying that information or what changes must be made to do so. Before changing a well-structured business process, you may find it necessary to educate the new leadership on the information that the current process is capable of providing. If the existing system was providing the data necessary to execute the business, it is likely capable of supplying most of the information after reorganization. In this way, you may be able to avoid wholesale changes to a solidly-designed program.

A final element of any continuous improvement process is the periodic review of the procedures governing the process. QA audits, periodic meetings with the user community, and meetings to address major changes should already be addressing the overall functionality of the data integrity. Therefore, the periodic review of your procedures should be focused on form and format. This approach may sound trivial. However, if users find obvious form and format errors in your documents, it is easy for them to consider the content outdated as well and avoid its use.

15.15 Sustainability is Not Optional

If we recognize that things that appear to be static data are not, the steps outlined above should help you avoid having your data and associated process become outdated. Business data is an asset in and of itself. The methods and processes that we have discussed above are the means by which you can maintain the integrity just as you maintain all the other assets that are critical to your business. For many, the journey to restoring your business data has been a long and often costly one. This has not been unlike restoring the health of your physical assets. And just like your physical assets, once you have taken the journey and paid the price, you recognize the importance of maintaining and protecting the investment you have made. You recognize that it takes some level of continued effort and investment to maintain your physical assets.

Taking the same approach and continuing to address and invest in your data is as important. Some would argue that perhaps it is more important than the investment in your physical assets. The continued health of your business data is, after all, part of the key to you understanding the health and condition of your physical assets. Without the health of your data, your ability to operate in an optimal manner is diminished to the point that you lack the ability to make value-added business decisions. In today's environment, sound data health is not optional. It is mandatory as the key to all your other business success.

Chapter 16

Getting Started

A journey of a thousand miles
begins with a single step.

Confucius

16.1 Getting Started

By now you realize that data integrity is serious business for everyone. Bad data results in bad decisions. Conversely, good, well-maintained data leads to good decisions. Not only good decisions for those using the data in the execution of the daily work, but good decisions through the entire business model all the way up to the office of the CEO. However, realizing the importance of data integrity does not always prompt action for a wide variety of reasons. When you think about it and do the analysis, none of the reasons for not beginning or not cleaning up what you currently have will stand up to a critical review. So, use this book, use your resources—both internally and those you can acquire externally—to get your effort going. You can't afford not to take this vitally important step.

When we first thought about a book focused on asset data integrity, we did research and learned that few if any books on the market address this topic. Although a great many address data integrity, *asset-related* data integrity seemed to be overlooked. We hope we have remedied this oversight. As you probably now recognize, this subject is very complex. Addressing it fully is not accomplished by the wave of a magic wand. Instead it requires commitment from the top to the bottom of the organization, not just during the clean-up phase, but also forever! We are not suggesting merely task-related changes. We are advocating a cultural change in which the values of the firm shift—from being willing to accept poor data quality and muddle through the decision making process to one where the data is accurate and value-added decisions can be made for the benefit of all.

If you search the Internet, attend conferences, or work through your peer network, you will discover firms who have successfully implemented a data integrity work process. You will find many firms, just like yours, who are considering starting the effort. And you will find many who have numerous excuses as to why they can't get the process moving. Which one do you want to be? Keep that question in mind the next time you are searching for some asset-related data and can not find it.

Our message is clear. Asset-related data integrity is serious business—serious enough that you can't afford to wait one more day. So, get started, and good luck on your journey!

Bob DiStefano
Steve Thomas

Bibliography

Belanger, Dennis. A Vision of Enterprise Reliability, Management Resources Group, Inc., 2009

Belanger, Dennis. Materials Management—It's Half the Battle, Management Resources Group, Inc., 2008

De Graff, Diane and Alex Whittlesey. Data Is an Asset, Management Resources Group, Inc., 2007

Deming, W. Edwards. Out of the Crisis, Massachusetts Institute of Technology, Center for Advanced Engineering Study, Cambridge, MA, 1986, p. 4

DiStefano, Robert. Turning Corporate Asset Management into Real Earnings per Share Growth, Management Resources Group, Inc. for SMRP Nashville, TN, 2008

DiStefano, Robert. Elevating Maintenance and Reliability Practices—The Financial Business Case, Management Resources Group, Inc. for UPTIME Magazine, 2008

Dittmar, Lee and Jane Griffin. IQ Improvement Starts Here, White paper downloaded from Deloitte web site: http://deloittereview.com

Drummond, T. and P. Woodlock. Data Quality: The First Step on the Path to Master Data Management, White paper downloaded from ZDNet web site: http://whitepapers.zdnet.com/abstract.

Environmental Protection Agency, 40 CFR Part 68. "Chemical Accident Prevention Provisions"

Environmental Protection Agency, 40 CFR Part 355. "Emergency Planning and Notification"

Kolbasuck-McGee, Marianne. Article, Information Week, January, 2007

Loshin, D. Master Data Management, Burlington, MA: Morgan Kauffman 2009

Management Resources Group, Inc. Data Integrity Is a Universal Issue.

Maximising Evidential Weight. (n.d.). Retrieved October 22, 2009 from http://www.open-tec.com/maximizing_evidential.html

Moubray, J. Reliability–Centered Maintenance RCM II (second edition), New York, NY: Industrial Press, Inc., 1997

Nuclear Energy Institute, NUMARC 93-01. "Industry Guideline for Monitoring the Effectiveness of Maintenance at Nuclear Power Plants"

Occupational Safety and Health Administration, 29 CFR 1010.119. "Process Safety Management of Highly Hazardous Chemicals"

Palmer, Richard (Doc). Maintenance Planning and Scheduling Handbook, McGraw-Hill, 2009

Pipino, L.L., Y.W. Lee, & R.Y. Wang. Data Quality Assessment, Communications of the ACM 2002, pp. 211-218

Schein, Edgar H. Organizational Culture and Leadership, 2e, San Francisco: Jossey Bass, 1992

Schein, Edgar H. The Corporate Culture Survival Guide, San Francisco: Jossey Bass, 1999

Strategic Maintenance Solutions, (2009), retrieved December 27, 2009, from http://www.automation.rockwell.com/solutions/maintenance/approaches.html

Thomas, Stephen J. Successfully Managing Change in Organizations: A User's Guide, New York: Industrial Press, 200

Thomas, Stephen J. Improving Maintenance and Reliability Through Cultural Change, New York: Industrial Press, 2005

Torrance, R. and K. Schmalz. Master Data Made Easy, SAP—Centric EAM 2008

Trillium Software (Division of Harte-Hanks). Solving the Source Data Problem with Automated Data Profiling.

U.S. Department of Labor's Bureau of Labor Statistics. Occupational Employment and Wage Estimates, May 2006

U.S. Food and Drug Administration. 21 CFR 110, 111, 210, 211, 225, 226, 606, Public Law 107–204, Sarbanes-Oxley Act of 2002

U.S. Nuclear Regulatory Commission. 10 CFR 50.65, "Requirements for Monitoring the Effectiveness of Maintenance at Nuclear Power Plants"

White, Todd. Master Records are Not Optional! Management Resources Group, Inc., 2008

Index